Asynchronous Android

Harness the power of multi-core mobile processors to build responsive Android applications

Steve Liles

BIRMINGHAM - MUMBAI

Asynchronous Android

First published: December 2013

Production Reference: 1171213

Published by Packt Publishing Ltd.
Livery Place
35 Livery Street
Birmingham B3 2PB, UK.

ISBN 978-1-78328-687-4

www.packtpub.com

Cover Image by Jarek Blaminsky (milak6@wp.pl)

Credits

Author
Steve Liles

Reviewers
David Bakin
Elie Abu Haydar
Hassan Makki
Matt Preston
Hélder Vasconcelos

Acquisition Editors
Harsha Bharwani
Kunal Parikh

Lead Technical Editor
Larissa Pinto

Technical Editors
Arwa Manasawala
Anand Singh

Proofreaders
Ameesha Green
Maria Gould

Copy Editors
Alisha Aranha
Mradula Hegde
Karuna Narayanan
Shambhavi Pai
Alfida Paiva
Lavina Pereira
Adithi Shetty

Project Coordinator
Jomin Varghese

Indexer
Hemangini Bari

Graphics
Yuvraj Mannari
Valentina D'silva

Production Coordinator
Manu Joseph

Cover Work
Manu Joseph

About the Author

Steve Liles is a self-confessed geek and has been an Android fan since the launch day of the G1. When he isn't at work building publishing systems and apps for newspapers and magazines, you'll find him tinkering with his own apps, building 3D printers, or playing RTS games. He is currently working with a start-up to build an advertising system that links the print and digital worlds using computer vision on Android and iOS devices.

I would like to sincerely thank the technical reviewers, who delivered their invaluable feedback gently and constructively. Without them, this book would be a pale shadow of the thing you now hold in your hands.

I once read that book acknowledgements are apologies to the people who have suffered. I must, therefore, unreservedly thank and apologize to my wife, not only for her patience and support during this project, but also through all the years and projects gone before.

About the Reviewers

David Bakin's first concurrent program was a Tektronix terminal emulator, written in assembly for an IMLAC PDS-1, which was a PDP-8-like machine with a GPU. He's written a lot of multiprocess and multithread concurrent programs and a number of tools for visualizing and debugging concurrent programs since then. He says computers now are orders of magnitude more powerful than the PDS-1, but concurrent programming hasn't gotten any easier.

His favorite languages are C++ and Haskell, and he prefers to use strong typing and functional programming techniques to write correct code. On the other hand, he also likes Smalltalk, Mathematica, and SIMD code in assembler because they're a lot of fun.

Elie Abu Haydar, born and raised in Beirut, has been interested in software development since his high school years. In 2006, he graduated from the American University of Science and Technology in Lebanon with a BS in Computer Science. As a software developer at KnowledgeView Ltd., Elie is seriously involved in every aspect of newsroom and publishing products where he uses Java to develop publishing solutions. These include backend web services and frontend web and mobile applications. Elie also contributes in the technical research and development of existing and future company projects.

Hassan Makki is a computer and communication engineer. He was born in 1979 in Lebanon and graduated in 2005. Since graduation, he has worked as a software engineer and has long-standing experience in C++, Java, and Android development. He began developing and managing Web and Android apps in 2011, and has developed around 30 apps related to news, music, sports, and advertisements.

Matt Preston is a professional software engineer with 14 years of experience in developing and maintaining a variety of systems for large international news and media organizations. He has worked on a range of low-latency/high-concurrency projects, ranging from mobile apps to distributed systems. Recently, he has been working on low-latency search and analytics using Elasticsearch.

I would like to thank my friend and colleague Steve Liles for this opportunity to let me review this book. We worked together for over 12 years and he's one of the best developers I have ever worked with. I learned a lot from him; I still continue to do so. Finally, thanks to Erica for all the tea; sorry about the late nights.

Hélder Vasconcelos has been a senior software engineer at Airtel ATN (Dublin, Ireland) since October 2012. He has extensive experience in designing and developing real-time/multithreaded Java and C/C++ applications for the telecommunications and aviation industry. Apart from his day-to-day job, for the last three years, he has been designing and developing native Android applications for Bearstouch Software and other third-party clients.

He graduated with a degree in Electronic & Telecommunications Engineering from the University of Aveiro in January 2006. He worked as a VoIP systems engineer at RedeRia Innovation (Aveiro, Portugal) from January 2006 to June 2007. He also worked as a software engineer at Outsoft/PT Inovação (Aveiro, Portugal) from October 2007 to October 2012.

Thanks to everyone involved in this project for your hard work and commitment, my awesome wife Tania for her love and support, and my parents and family for their awesome effort in my education. Additionally, I would like to thank my friends, colleagues, clients, and teachers for helping me to shape and improve my skills and perspectives during my career.

www.PacktPub.com

Support files, eBooks, discount offers and more

You might want to visit www.PacktPub.com for support files and downloads related to your book.

Did you know that Packt offers eBook versions of every book published, with PDF and ePub files available? You can upgrade to the eBook version at www.PacktPub.com and as a print book customer, you are entitled to a discount on the eBook copy. Get in touch with us at service@packtpub.com for more details.

At www.PacktPub.com, you can also read a collection of free technical articles, sign up for a range of free newsletters and receive exclusive discounts and offers on Packt books and eBooks.

http://PacktLib.PacktPub.com

Do you need instant solutions to your IT questions? PacktLib is Packt's online digital book library. Here, you can access, read and search across Packt's entire library of books.

Why Subscribe?
- Fully searchable across every book published by Packt
- Copy and paste, print and bookmark content
- On demand and accessible via web browser

Free Access for Packt account holders

If you have an account with Packt at www.PacktPub.com, you can use this to access PacktLib today and view nine entirely free books. Simply use your login credentials for immediate access.

Table of Contents

Preface

Programming is the most fun a person can have on their own. This is a fact well-known to programmers, though it seems the rest of the world is yet to catch on. You already know this or you wouldn't be reading this book, but it constantly amazes me that more people aren't falling over themselves to learn to code.

Meanwhile, mobile devices have made computers fun even for non-coders. We carry in our pockets small machines with incredible processing power and a giddying array of sensors and interfaces.

Android takes these fun machines and makes them accessible to programmers through a fabulously well-crafted platform and tool chain, in a programming language that has stood the test of time yet continues to develop and evolve.

What could possibly be better than programming fun machines to do cool things, in a powerful language, on a well-crafted platform, with a world-class tool chain? For me, the answer is doing so with a good enough understanding of those things to make the difference between a good app and a great one.

There are many things that must come together to make a great app. You need a great idea—I can't help you there. You need a pretty user interface—sorry, wrong book. You need a great user experience—aha! Now we're getting somewhere. Among the many things that contribute to a great user experience, responsiveness is right up there near the top of the list.

It's easiest to define responsiveness with examples of its lack: pauses and glitches while scrolling content, user interfaces that freeze while loading data from storage, applications that don't give progress updates to let us know what's happening, failing to complete work that we initiated, staring at a spinner while data is fetched from the network, and the list goes on.

This book is about making the difference between a good app and a great one; smoothing out the glitches, keeping the UI responsive, telling the user how things are going, making sure we finish what we started, using those powerful multicore processors, and doing it all without wasting the battery. Let's have some fun!

What this book covers

Chapter 1, Building Responsive Android Applications, gives an overview of the Android process and thread model, and describes some of the challenges and benefits of concurrency in general, before discussing issues specific to Android.

Chapter 2, Staying Responsive with AsyncTask, covers the poster child of concurrent programming in Android. We learn how AsyncTask works, how to use it correctly, and how to avoid the common pitfalls that catch out even experienced developers.

Chapter 3, Distributing Work with Handler and HandlerThread, details the fundamental and related topics of Handler, HandlerThread, and Looper, and illustrates how they can be used to schedule tasks on the main thread, and to coordinate and communicate work between cooperating background threads.

Chapter 4, Asynchronous I/O with Loader, introduces the Loader framework and tackles the important task of loading data asynchronously to keep the user interface responsive and glitch free.

Chapter 5, Queuing Work with IntentService, gives us the means to perform background operations beyond the scope of a single Activity lifecycle and to ensure that our work is completed even if the user leaves the application.

Chapter 6, Long-running Tasks with Service, extends the capabilities we discovered with IntentService and gives us control over the level of concurrency applied to our long-running background tasks.

Chapter 7, Scheduling Work with AlarmManager, completes our toolkit by enabling us to arrange for work to be done far into the future and on repeating schedules. It also enables us to build apps that alert users to new content and start instantly with fresh data.

What you need for this book

To follow along and experiment with the examples, you will need a development computer with a Java 6 (or 7) SE Development Kit and the Android Software Development Kit Version 7 or above (you will need at least Version 19 to try all of the examples).

You will also need an Integrated Development Environment such as Android Studio or Eclipse. The examples have been developed using Google's new Android Studio IDE and use its integrated build system, Gradle.

While you can run the examples using the emulator provided by the Android SDK, it is a poor substitute for the real thing. A physical Android device is a much faster and more pleasurable way to develop and test Android applications!

Many of the examples will work on a device running any version of Android since 2.1, Éclair. Some examples demonstrate newer APIs and as a result, require a more recent Android version—up to Android 4.4, KitKat.

You can also download a prebuilt application containing all of the examples from Google Play; search for "Asynchronous Android".

Who this book is for

This book is for developers who have mastered the basics of Android and are ready to take the next big step to improve the quality of your apps—not just behind the scenes engineering quality, but real perceivable improvements that make a difference to end users.

A reasonable understanding of core Android development is assumed. If you have built Android apps before and are comfortable with the Activity class and its lifecycle, XML layout files, and the Android manifest, you should have no problem understanding the topics in this book.

Familiarity with Java's concurrency primitives and higher-level constructs will aid and deepen understanding, but is not a prerequisite.

Android developers with no prior experience of concurrency and asynchronous programming will learn when, why, and how to apply Android's concurrency constructs to build responsive apps.

Java experts new to Android will be equipped to properly apply their existing knowledge in the Android environment and will discover elegant solutions to familiar problems in Android's high-level concurrency constructs.

Conventions

In this book, you will find a number of styles of text that distinguish between different kinds of information. Here are some examples of these styles, and an explanation of their meaning.

Code words in text are shown as follows: "We can include other contexts through the use of the include directive."

A block of code is set as follows:

```
protected void onPause() {
    super.onPause();
    if (task != null)
        task.cancel(false);
}
```

When we wish to draw your attention to a particular part of a code block, the relevant lines or items are set in bold:

```
protected void onPause() {
    super.onPause();
    if (task != null)
        task.cancel(false);
}
```

Any command-line input or output is written as follows:

```
> telnet 192.168.0.4 4444
```

New terms and **important words** are shown in bold. Words that you see on the screen, in menus or dialog boxes for example, appear in the text like this: "clicking on the **Next** button moves you to the next screen."

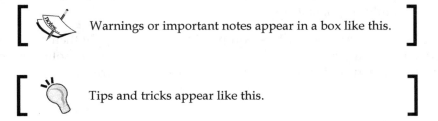

Warnings or important notes appear in a box like this.

Tips and tricks appear like this.

Reader feedback

Feedback from our readers is always welcome. Let us know what you think about this book—what you liked or may have disliked. Reader feedback is important for us to develop titles that you really get the most out of.

To send us general feedback, simply send an e-mail to feedback@packtpub.com, and mention the book title via the subject of your message.

If there is a topic that you have expertise in and you are interested in either writing or contributing to a book, see our author guide on www.packtpub.com/authors.

Customer support

Now that you are the proud owner of a Packt book, we have a number of things to help you to get the most from your purchase.

Downloading the example code

You can download the example code files for all Packt books you have purchased from your account at http://www.packtpub.com. If you purchased this book elsewhere, you can visit http://www.packtpub.com/support and register to have the files e-mailed directly to you.

Errata

Although we have taken every care to ensure the accuracy of our content, mistakes do happen. If you find a mistake in one of our books—maybe a mistake in the text or the code—we would be grateful if you would report this to us. By doing so, you can save other readers from frustration and help us improve subsequent versions of this book. If you find any errata, please report them by visiting http://www.packtpub.com/submit-errata, selecting your book, clicking on the **errata submission form** link, and entering the details of your errata. Once your errata are verified, your submission will be accepted and the errata will be uploaded on our website, or added to any list of existing errata, under the Errata section of that title. Any existing errata can be viewed by selecting your title from http://www.packtpub.com/support.

Piracy

Piracy of copyright material on the Internet is an ongoing problem across all media. At Packt, we take the protection of our copyright and licenses very seriously. If you come across any illegal copies of our works, in any form, on the Internet, please provide us with the location address or website name immediately so that we can pursue a remedy.

Please contact us at copyright@packtpub.com with a link to the suspected pirated material.

We appreciate your help in protecting our authors, and our ability to bring you valuable content.

Questions

You can contact us at questions@packtpub.com if you are having a problem with any aspect of the book, and we will do our best to address it.

1
Building Responsive Android Applications

The Android operating system has, at its heart, a heavily modified Linux kernel designed to securely and efficiently run many process virtual machines on devices with relatively limited resources.

To build Android applications that run smoothly and responsively in these resource-constrained environments, we need to arm ourselves with an understanding of the options available, and how, when, and why to use them — this is the essence of this book.

However, before we do that, we'll briefly consider why we need to concern ourselves at all. We'll see how serious Google is about the efficiency of the platform, explore the Android process model and its implications for programmers and end users, and examine some of the measures that the Android team have put in place to protect users from apps that behave badly.

To conclude, we'll discuss the general approach used throughout the rest of the book to keep applications responsive using asynchronous programming and concurrency, and its associated challenges and benefits.

In this chapter, we will cover the following topics:

- Introducing the Dalvik Virtual Machine
- Memory sharing and the Zygote
- Understanding the Android thread model
- The main thread

- Unresponsive apps and the ANR dialog
- Maintaining responsiveness
- Concurrency in Android

Introducing the Dalvik Virtual Machine

Android applications are typically programmed using the Java language, but the virtual machines in the Android stack are not instances of the **Java Virtual Machine (JVM)**. Instead, the Java source is compiled to Java byte-code and translated into a **Dalvik executable file (DEX)** for execution on a **Dalvik Virtual Machine (DVM)**.

It is no accident that Google chose Java as the primary language, allowing a vast pool of developer talent to quickly get to work on building apps, but why not simply run Android applications directly on a JVM?

Dalvik was created specifically for Android, and as such, was designed to operate in environments where memory, processor, and electrical power are limited, for example, mobile devices. Satisfying these design constraints resulted in a very different virtual machine than the typical JVM's that we know from desktop and server environments.

Dalvik goes to great lengths to improve the efficiency of the JVM, involving a range of optimizations to simplify and speed up interpretation and reduce the memory footprint of a running program. The most fundamental difference between the two VM architectures is that the JVM is a stack-based machine, whereas the DVM is register-based.

A stack-based virtual machine must transfer data from registers to the operand stack before manipulating them. In contrast, a register-based VM operates by directly using virtual registers. This increases the relative size of instructions because they must specify which registers to use, but reduces the number of instructions that must be executed to achieve the same result.

Dalvik's creators claim that the net result is in Dalvik's favour and that the DVM is on average around 30 percent more efficient than the JVM. Clearly, Google has gone to great lengths to squeeze every last drop of performance out of each mobile device to help developers build responsive applications!

Memory sharing and the Zygote

Another huge efficiency of the platform is brought about by the way in which a new DVM instance is created and managed.

A special process called the **Zygote** is launched when Android initially boots. The Zygote starts up a virtual machine, preloads the core libraries, and initializes various shared structures. It then waits for instructions by listening on a socket.

When a new Android application is launched, the Zygote receives a command to create a virtual machine to run the application on. It does this by forking its prewarmed VM process and creating a new child process that shares its memory with the parent, using a technique called **Copy-On-Write**. This has some fantastic benefits:

- First, the virtual machine and core libraries are already loaded into the memory. Not having to read this significant chunk of data to initialize the virtual machine drastically reduces the startup overhead.

- Second, the memory in which these core libraries and common structures reside is shared by the Zygote with all other applications, resulting in saving a lot of memory when the user is running multiple apps.

Understanding the Android thread model

Each forked application process runs independently and is scheduled frequent, small amounts of CPU time by the operating system. This time-slicing approach means that even a single-processor device can appear to be actively working in more than one application at the same time, when in fact, each application is taking very short turns on the CPU.

Within a process, there may be many threads of execution. Each thread is a separate sequential flow of control within the overall program — it executes its instructions in order, one after the other. These threads are also allocated slices of CPU time by the operating system.

While the application process is started by the system and prevented from directly interfering with data in the memory address space of other processes, threads may be started by application code and can communicate and share data with other threads within the same process.

The main thread

Within each DVM process, the system starts a number of threads to perform important duties such as garbage collection, but of particular importance to application developers is the single thread of execution known as the main or UI thread. By default, any code that we write in our applications will be executed by the main thread.

For example, when we write code in an onCreate method in the Activity class, it will be executed on the main thread. Likewise, when we attach listeners to user-interface components to handle taps and other user-input gestures, the listener callback executes on the main thread.

For apps that do little I/O or processing, this single thread model is fine. However, if we need to do CPU-intensive calculations, read or write files from permanent storage, or talk to a web service, any further events that arrive while we're doing this work will be blocked until we're finished.

Unresponsive apps and the ANR dialog

As you can imagine, if the main thread is busy with a heavy calculation or reading data from a network socket, it cannot immediately respond to user input such as a tap or swipe.

An app that doesn't respond quickly to user interaction will feel unresponsive — anything more than a couple of hundred milliseconds delay is noticeable. This is such a pernicious problem that the Android platform protects users from applications that do too much on the main thread.

> If an app does not respond to user input within 5 seconds, the user will see the **Application Not Responding** (ANR) dialog and be offered the option to quit the app.

Android works hard to synchronize user interface redraws with the hardware refresh rate. This means that it aims to redraw at 60 frames per second — that's just 16.67 ms per frame. If we do work on the main thread that takes anywhere near 16 ms, we risk affecting the frame rate, resulting in jank — stuttering animations, jerky scrolling, and so on.

At API level 16, Android introduced a new entity, the Choreographer, to oversee timing issues. It will start issuing dropped-frame warnings in the log if you drop more than 30 consecutive frames.

Ideally, of course, we don't want to drop a single frame. Jank, unresponsiveness, and especially the ANR, offer a very poor user experience and translate into bad reviews and an unpopular application. A rule to live by when building Android applications is: do not block the main thread!

 Android provides a helpful strict mode setting in **Developer Options** on each device, which will flash the screen when applications perform long-running operations on the main thread.

Further protection was added to the platform in Honeycomb (API level 11) with the introduction of a new Exception class, NetworkOnMainThreadException, a subclass of RuntimeException that is thrown if the system detects network activity initiated on the main thread.

Maintaining responsiveness

Ideally then, we want to offload long-running operations from the main thread so that they can be handled in the background, and the main thread can continue to process user interface updates smoothly and respond in a timely fashion to user interaction.

For this to be useful, we must be able to coordinate work and safely pass data between cooperating threads—especially between background threads and the main thread.

We also want to execute many background tasks at the same time and take advantage of additional CPU cores to churn through heavy processing tasks quickly.

This simultaneous execution of separate code paths is known as **concurrency**.

Concurrency in Android

The low-level constructs of concurrency in Android are those provided by the Java language: java.lang.Thread, java.lang.Runnable, and the synchronized and volatile keywords.

Higher-level mechanisms introduced to Java 5 in the java.util.concurrent package, such as Executors, atomic wrapper classes, locking constructs, and concurrent collections, are also available for use in Android applications.

We can start new threads of execution in our Android applications just as we would in any other Java application, and the operating system will schedule some CPU time for those threads.

To do some work off the main thread, we can simply create a new instance of `java.lang.Thread`, override its `run()` method with the code we want it to execute, and invoke its `start()` method.

While starting new threads is easy, concurrency is actually a very difficult thing to do well. Concurrent software faces many issues that fall into the two broad categories: correctness (producing consistent and correct results) and liveness (making progress towards completion).

Correctness issues in concurrent programs

A common example of a correctness problem occurs when two threads need to modify the value of the same variable based on its current value. Let's imagine that we have an integer variable `myInt` with the current value of 2.

In order to increment `myInt`, we first need to read its current value and then add 1 to it. In a single threaded world, the two increments would happen in a strict sequence — we read the initial value 2, add 1 to it, set the new value back to the variable, then repeat the sequence. After the two increments, `myInt` holds the value 4.

In a multithreaded environment, we run into potential timing issues. It is possible that two threads trying to increment the variable would both read the same initial value (2), add 1 to it, and set the result (in both cases, 3) back to the variable.

Both threads have behaved correctly in their localized view of the world, but in terms of the overall program, we clearly have a correctness problem; 2 + 2 should not equal 3! This kind of timing issue is known as a **race condition**.

A common solution to correctness problems such as race conditions is **mutual exclusion** — preventing multiple threads from accessing certain resources at the same time. Typically, this is achieved by ensuring that threads acquire an exclusive lock before reading or updating shared data.

Liveness issues in concurrent programs

Liveness can be thought of as the ability of the application to do useful work and make progress towards goals. Liveness problems tend to be an unfortunate side effect of the solution to correctness problems. By locking access to data or system resources, it is possible to create bottlenecks where many threads are contending for access to a single lock, leading to potentially significant delays.

Worse, where multiple locks are used, it is possible to create a situation where no thread can make progress because each requires exclusive access to a lock that another thread currently owns—a situation known as a **deadlock**.

Android-specific concurrency issues

There are two additional problems facing developers of concurrent Android applications which are specific to Android.

The Activity lifecycle

Android applications are typically composed of one or more subclasses of android. app.Activity. An Activity instance has a very well-defined lifecycle that the system manages through the execution of lifecycle method callbacks, all of which are executed on the main thread.

An Activity instance that has been completed should be eligible for garbage collection, but background threads that refer to the Activity or part of its view hierarchy can prevent garbage collection and create a memory leak.

Similarly, it is easy to waste CPU cycles (and battery life) by continuing to do background work when the result can never be displayed because Activity has finished.

Finally, the Android platform is free at any time to kill processes that are not the user's current focus. This means that if we have long-running operations to complete, we need some way of letting the system know not to kill our process yet!

All of this complicates the do-not-block–the-main-thread rule because we need to worry about canceling background work in a timely fashion or decoupling it from the Activity lifecycle where appropriate.

Manipulating the user interface

The other Android-specific problem lies not in what you can do from the UI thread, but in what you cannot do:

 You cannot manipulate the user interface from any thread other than the main thread.

This is because the user-interface toolkit is not thread-safe, that is, accessing it from multiple threads may cause correctness problems. In fact, the user-interface toolkit protects itself from potential problems by actively denying access to user-interface components from threads other than the one that originally created those components.

The final challenge then lies in safely synchronizing background threads with the main thread so that the main thread can update the user interface with the results of the background work.

Android-specific concurrency constructs

The good news is that the Android platform provides specific constructs to address the general issues of concurrency, and to solve the specific problems presented by Android.

There are constructs that allow us to defer tasks to run later on the main thread, to communicate easily between cooperating threads, and to issue work to managed pools of worker threads and re-integrate the results.

There are solutions to the constraints of the Activity lifecycle both for medium-term operations that closely involve the user-interface and for longer-term work that must be completed even if the user leaves the application.

While some of these constructs were only introduced with newer releases of the Android platform, all are available through the support libraries and, with a few exceptions, the examples in this book target devices that run API level 7 (Android 2.1) and above.

The rest of this book discusses these Android-specific constructs and their usage and applications.

Summary

In this chapter, we learned that Google takes the efficiency of the Android platform very seriously. We also looked at the extraordinary lengths they go to in order to ensure a smooth user experience, evidencing the importance of building responsive applications.

We discussed the Android thread model and the measures that the platform may take to protect the user from apps that misbehave or are not sufficiently responsive.

Finally, we gained an overview of the general approach to building responsive apps through concurrency, and learned some of the issues faced by developers of concurrent software in general and Android applications in particular.

In the next chapter, we'll start to build responsive applications by applying the infamous AsyncTask instance to execute work in the background using pools of threads, and return progress updates and results to the main thread.

2
Staying Responsive with AsyncTask

The first Android-specific concurrency construct we'll look at is `android.os.AsyncTask`, a neat construct that encapsulates the messy business of managing threads, performing background work, and publishing progress and results back to the main thread to update the user interface.

In this chapter we will cover the following topics:

- Introducing `AsyncTask`
- Declaring `AsyncTask` types
- Executing AsyncTasks
- Providing feedback to the user
- Providing progress updates
- Canceling AsyncTasks
- Handling exceptions
- Controlling the level of concurrency
- Common `AsyncTask` issues
- Applications of `AsyncTask`

Introducing AsyncTask

AsyncTask was introduced in Android at API level 3, Cupcake, with the express purpose of helping developers to avoid blocking the main thread. The Async part of the name of this class comes from the word asynchronous, which literally means not occurring at the same time.

AsyncTask is an abstract class, and as such, must be subclassed for use. At the minimum, our subclass must provide an implementation for the abstract doInBackground method, which defines the work that we want to get done off the main thread.

```
protected Result doInBackground(Params... params)
```

There are four other methods of AsyncTask which we may choose to override:

```
protected void onPreExecute()
protected void onProgressUpdate(Progress... values)
protected void onPostExecute(Result result)
protected void onCancelled(Result result)
```

Although we will override one or more of these five methods, we will *not* invoke them directly from our own code. These are **callback** methods, meaning that they will be invoked for us (called back) at the appropriate time.

The key difference between doInBackground and the other four methods is the thread on which they execute.

Before any background work begins, onPreExecute will be invoked and will run to completion on the main thread.

Once onPreExecute completes, doInBackground will be scheduled and will start work on a background thread.

During the background work, we can publish progress updates from doInBackground, which trigger the main thread to execute onProgressUpdate with the progress values we provide. By invoking this on the main thread, AsyncTask makes it easy for us to update the user interface to show progress (remember that we can only update the user interface from the main thread).

When the background work completes successfully, doInBackground may return a result. This result is passed to onPostExecute, which is invoked for us on the main thread so that we can update the user interface with the results of our background processing.

This pattern of passing data from one thread to another is very important, because it helps us to avoid several thread-safety issues.

Our AsyncTask could manipulate fields of the enclosing Activity class, but then we would have to take extra precautions, such as adding synchronization to prevent race conditions and ensure visibility of updates.

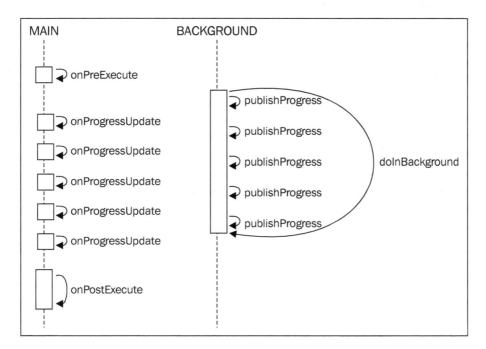

The preceding figure displays a sequence of method calls executed by AsyncTask, illustrating which methods run on the main thread versus the AsyncTask background thread.

If we invoke AsyncTask's cancel method before doInBackground completes, onPostExecute will not be called. Instead, the alternative onCancelled callback method is invoked so that we can implement different behavior for a successful versus cancelled completion.

 The onPreExecute, onProgressUpdate, onPostExecute, and onCancelled methods are invoked on the main thread, so we must not perform long-running/blocking operations in these methods.

Declaring AsyncTask types

AsyncTask is a generically typed class, and exposes three type parameters:

```
abstract class AsyncTask<Params, Progress, Result>
```

When we declare an AsyncTask subclass, we'll specify types for Params, Progress, and Result; for example, if we want to pass a String parameter to doInBackground, report progress as a Float, and return a Boolean result, we would declare our AsyncTask subclass as follows:

```
public class MyTask extends AsyncTask<String, Float, Boolean>
```

If we don't need to pass any parameters, or don't want to report progress, a good type to use for those parameters is java.lang.Void, which signals our intent clearly, because Void is an uninstantiable class representing the void keyword.

Let's take a look at a first example, performing an expensive calculation in the background and reporting the result to the main thread:

```
public class PrimesTask
extends AsyncTask<Integer, Void, BigInteger> {
  private TextView resultView;

  public PrimesTask(TextView resultView) {
    this.resultView = resultView;
  }

  @Override
  protected BigInteger doInBackground(Integer... params) {
    int n = params[0];
    BigInteger prime = new BigInteger("2");
    for (int i=0; i<n; i++) {
      prime = prime.nextProbablePrime();
    }
    return prime;
  }

  @Override
  protected void onPostExecute(BigInteger result) {
      resultView.setText(result.toString());
  }
}
```

Here, `PrimesTask` extends `AsyncTask`, specifying the `Params` type as `Integer` so that we can ask for the *n*th prime, and the `Result` type as `BigInteger`.

We pass a `TextView` to the constructor so that `PrimesTask` has a reference to the user interface that it should update upon completion.

We've implemented `doInBackground` to calculate the *n*th prime, where *n* is an `Integer` parameter to `doInBackground`, and returned the result as `BigInteger`.

In `onPostExecute`, we simply display the `result` parameter to the view we were assigned in the constructor.

Downloading the example code

You can download the example code files for all Packt books you have purchased from your account at http://www.packtpub.com. If you purchased this book elsewhere, you can visit http://www.packtpub.com/support and register to have the files e-mailed directly to you.

Executing AsyncTasks

Having implemented `doInBackground` and `onPostExecute`, we want to get our task running. There are two methods we can use for this, each offering different levels of control over the degree of concurrency with which our tasks are executed. Let's look at the simpler of the two methods first:

```
public final AsyncTask<Params, Progress, Result> execute(Params…
params)
```

The return type is the type of our `AsyncTask` subclass, which is simply for convenience so that we can use method chaining to instantiate and start a task in a single line and still record a reference to the instance:

```
class MyTask implements AsyncTask<String,Void,String>{ … }
MyTask task = new MyTask().execute("hello");
```

The `Params… params` argument is the same `Params` type we used in our class declaration, because the values we supply to the `execute` method are later passed to our `doInBackground` method as its `Params… params` arguments. Notice that it is a `varargs` parameter, meaning that we can pass any number of parameters of that type (including none).

> Each instance of AsyncTask is a single-use object — once we have
> started an AsyncTask, it can never be started again, even if we
> cancel it or wait for it to complete first.
>
> This is a safety feature, designed to protect us from concurrency
> issues such as the race condition we saw in *Chapter 1, Building
> Responsive Android Applications*.

Executing PrimesTask is straightforward — we need Activity, which constructs an
instance of PrimesTask with a view to update, then invokes execute with a suitable
value for n:

```java
public class PrimesActivity extends Activity {
  @Override
  protected void onCreate(Bundle savedInstanceState) {
    super.onCreate(savedInstanceState);

    setContentView(R.layout.activity_example_1);
    final TextView resultView =
      (TextView) findViewById(R.id.result);
    Button goButton = (Button) findViewById(R.id.go);

    goButton.setOnClickListener(
      new View.OnClickListener() {
        @Override
          public void onClick(View view) {
            new PrimesTask(resultView).execute(500);
      }
    });
  }
}
```

Great! We're not blocking up the main thread, so our app is nice and responsive,
but we can do better.

Providing feedback to the user

Having started what we know to be a potentially long-running task, we probably want
to let the user know that something is happening. There are a lot of ways of doing this,
but a common approach is to present a dialog displaying a relevant message.

A good place to present our dialog is from the onPreExecute method of AsyncTask, which executes on the main thread. Hence, it is allowed to interact with the user interface.

The modified PrimesTask will need a reference to a Context, so that it can prepare a ProgressDialog, which it will show and dismiss in onPreExecute and onPostExecute respectively. As doInBackground has not changed, it is not shown in the following code, for brevity:

```
public class PrimesTask extends AsyncTask<Integer, Void, BigInteger>{
  private Context ctx;
  private ProgressDialog progress;
  private TextView resultView;

  public PrimesTask(Context ctx, TextView resultView) {
    this.ctx = ctx;
      this.resultView = resultView;
  }

  @Override
  protected void onPreExecute() {
    progress = new ProgressDialog(ctx);
    progress.setTitle(R.string.calculating);
    progress.setCancelable(false);
    progress.show();
  }

  // … doInBackground elided for brevity …

  @Override
  protected void onPostExecute(BigInteger result) {
    resultView.setText(result.toString());
    progress.dismiss();
  }
}
```

All that remains is to pass a Context to the constructor of our modified PrimesTask. As Activity is a subclass of Context, we can simply pass a reference to the host Activity:

```
goButton.setOnClickListener(new View.OnClickListener() {
  public void onClick(View view) {
    new PrimesTask(
      PrimesActivity.this, resultView).execute(500);
  }
});
```

Providing progress updates

Knowing that something is happening is a great relief to our users, but they might be getting impatient and wondering how much longer they need to wait. Let's show them how we're getting on by adding a progress bar to our dialog.

Remember that we aren't allowed to update the user interface directly from `doInBackground`, because we aren't on the main thread. How, then, can we tell the main thread to make these updates for us?

`AsyncTask` comes with a handy callback method for this, whose signature we saw at the beginning of the chapter:

```
protected void onProgressUpdate(Progress... values)
```

We can override `onProgressUpdate` to update the user interface from the main thread, but when does it get called and where does it get its `Progress...` values from? The glue between `doInBackground` and `onProgressUpdate` is another of AsyncTask's methods:

```
protected final void publishProgress(Progress... values)
```

To update the user interface with our progress, we simply *publish* progress updates from the background thread by invoking `publishProgress` from within `doInBackground`. Each time we call `publishProgress`, the main thread will be scheduled to invoke `onProgressUpdate` for us with these progress values.

The modifications to our running example to show a progress bar are quite simple. First, we must change the class declaration to include a `Progress` type. We'll be setting progress values in the range `0` to `100`, so we'll use `Integer`:

```
public class PrimesTask
    extends AsyncTask<Integer, Integer, BigInteger> {
```

Next, we need to set the style and the bounds of the progress bar. We can do that with the following additions to `onPreExecute`:

```
progress.setProgressStyle(ProgressDialog.STYLE_HORIZONTAL);
progress.setProgress(0);
progress.setMax(100);
```

We also need to implement the `onProgressUpdate` callback to update the progress bar from the main thread:

```
@Override
protected void onProgressUpdate(Integer... values) {
  progress.setProgress(values[0]);
}
```

The final modification is to calculate the progress at each iteration of the `for` loop, and invoke `publishProgress` so that the main thread knows to call back `onProgressUpdate`:

```
protected BigInteger doInBackground(Integer... params) {
  int primeToFind = params[0];
  BigInteger prime = new BigInteger("2");
  int progress = 0;
  for (int i=0; i<primeToFind; i++) {
    prime = prime.nextProbablePrime();
    int percent = (int)((i * 100f)/primeToFind);
    if (percent > progress) {
      publishProgress(percent);
      progress = percent;
    }
  }
  return prime;
}
```

It is important to understand that invoking `publishProgress` does not directly invoke the main thread, but adds a task to the main thread's queue, which will be processed at some time in the near future by the main thread.

Notice that we're being careful to publish progress only when the percentage actually changes, avoiding any unnecessary overhead.

The delay between publishing the progress and seeing the user interface update will be extremely short, and the progress bar will update smoothly, provided we are careful to follow the golden rule of not blocking the main thread from any of our code.

Canceling AsyncTask

Another nice usability touch we can provide for our users is the ability to cancel a task before it completes — for example, if the task depends on some user input and, after starting the execution, the user realizes that they have provided the wrong value. AsyncTask provides support for cancellation with the `cancel` method.

```
public final boolean cancel(boolean mayInterruptIfRunning)
```

The `mayInterruptIfRunning` parameter allows us to specify whether an `AsyncTask` thread that is in an interruptible state may actually be interrupted — for example, if our `doInBackground` code is performing interruptible I/O.

Simply invoking `cancel` is not sufficient to cause our task to finish early. We need to actively support cancellation by periodically checking the value returned from `isCancelled` and reacting appropriately in `doInBackground`.

First, let's set up our `ProgressDialog` to trigger the AsyncTask's `cancel` method by adding a few lines to `onPreExecute`:

```
progress.setCancelable(true);
progress.setOnCancelListener(
  new DialogInterface.OnCancelListener() {
    public void onCancel(DialogInterface dialog) {
      PrimesTask.this.cancel(false);
    }
});
```

Now we can trigger `cancel` by touching outside the progress dialog, or pressing the device's back button while the dialog is visible. We'll invoke `cancel` with `false`, as we are not doing interruptible work inside the method or checking the return value of `Thread.interrupted`, so calling an interrupt will have no effect. We still need to check for cancellation in `doInBackground`, so we will modify it as follows:

```
protected BigInteger doInBackground(Integer... params) {
  int primeToFind = params[0];
  BigInteger prime = new BigInteger("2");
  for (int i=0; i<primeToFind; i++) {
    prime = prime.nextProbablePrime();
    int percentComplete = (int)((i * 100f)/primeToFind);
    publishProgress(percentComplete);
    if (isCancelled())
      break;
  }
  return prime;
}
```

The cancelled `AsyncTask` does not receive the `onPostExecute` callback. Instead, we have the opportunity to implement different behavior for a cancelled execution by implementing `onCancelled`. There are two variants of this callback method:

```
protected void onCancelled(Result result);
protected void onCancelled();
```

The default implementation of the parameterized `onCancelled(Result result)` method delegates to the `onCancelled` method.

If `AsyncTask` can provide either a complete result (such as a fully downloaded image) or nothing, then we will probably want to override the zero argument `onCancelled()` method.

If we are performing an incremental computation in our `AsyncTask`, we might choose to override the `onCancelled(Result result)` version so that we can make use of the result computed up to the point of cancellation.

In all cases, since `onPostExecute` does not get called on a canceled `AsyncTask`, we will want to make sure that our `onCancelled` callbacks update the user interface appropriately — in our example, this entails dismissing the progress dialog we opened in `onPreExecute`, and updating the result text.

```
protected void onCancelled(BigInteger result) {
  if (result != null)
    resultView.setText("cancelled at " + result.toString());
  progress.dismiss();
}
```

Another situation to be aware of occurs when we cancel an `AsyncTask` that has not yet begun its `doInBackground` method. If this happens, `doInBackground` will never be invoked, though `onCancelled` will still be called on the main thread.

Handling exceptions

The callback methods defined by `AsyncTask` dictate that we cannot throw checked exceptions, so we must wrap any code that throws checked exceptions with `try/catch` blocks. Unchecked exceptions that propagate out of AsyncTask's methods will crash our application, so we must test carefully and handle these if necessary.

For the callback methods that run on the main thread — `onPreExecute`, `onProgressUpdate`, `onPostExecute`, and `onCancelled` — we can catch exceptions in the method and directly update the user interface to alert the user.

Of course, exceptions are likely to arise in our `doInBackground` method too, as this is where the bulk of the work of `AsyncTask` is done, but unfortunately, we can't update the user interface from `doInBackground`. A simple solution is to have `doInBackground` return an object that may contain either the result or an exception, as follows:

```
static class Result<T> {
  private T actual;
  private Exception exc;
}
```

```
@Override
protected final Result<T> doInBackground(Void... params) {
  Result<T> result = new Result<T>();
  try {
    result.actual = calculateResult();
  } catch (Exception exc) {
    result.exc = exc;
  }
    return result;
}


protected abstract T calculateResult() throws Exception;
```

Now we can check in onPostExecute for the presence of an Exception in the Result object. If there is one, we can deal with it, perhaps by alerting the user; otherwise, we just use the actual result as normal.

```
@Override
protected final void onPostExecute(Result<R> result) {
    if (result.exc != null) {
        // … alert the user …
    } else {
        // … success, continue as normal …
    }
}
```

In the downloadable code, you'll find a handy SafeAsyncTask class that takes care of exception handling, and makes it easy to deal with exceptions or results.

Controlling the level of concurrency

So far, we've carefully avoided being too specific about what exactly happens when we invoke AsyncTask's execute method. We know that doInBackground will execute off the main thread, but what exactly does that mean?

The original goal of AsyncTask was to help developers avoid blocking the main thread. In its initial form at API level 3, AsyncTasks were queued and executed serially (that is, one after the other) on a single background thread, guaranteeing that they would complete in the order they were started.

This changed in API level 4 to use a pool of up to 128 threads to execute multiple AsyncTasks concurrently with each other—a level of concurrency of up to 128. At first glance, this seems like a good thing, since a common use case for `AsyncTask` is to perform blocking I/O, where the thread spends much of its time idly waiting for data.

However, as we saw in *Chapter 1, Building Responsive Android Applications*, there are many issues that commonly arise in concurrent programming, and indeed, the Android team realized that by executing AsyncTasks concurrently by default, they were exposing developers to potential programming problems (for example, when executed concurrently, there are no guarantees that AsyncTasks will complete in the same order they were started).

As a result, a further change was made at API level 11, switching back to serial execution by default, and introducing a new method that gives concurrency control back to the app developer:

```
public final AsyncTask<Params, Progress, Result>
executeOnExecutor(Executor exec, Params... params)
```

From API level 11 onwards, we can start AsyncTasks with `executeOnExecutor`, and in doing so, choose the level of concurrency for ourselves by supplying an `Executor` object.

`Executor` is an interface from the `java.util.concurrent` package of the JDK, and was first introduced in Java 5. Its purpose is to present a way to submit tasks for execution without spelling out precisely how or when the execution will be carried out.

Implementations of `Executor` may run tasks sequentially using a single thread, use a limited pool of threads to control the level of concurrency, or even directly create a new thread for each task.

The `AsyncTask` class provides two `Executor` interfaces that allow you to choose between the concurrency levels described earlier in this section:

- `SERIAL_EXECUTOR`: This `Executor` queues tasks and uses a single background thread to run them to completion, each in turn in the order they were submitted.
- `THREAD_POOL_EXECUTOR`: This `Executor` runs tasks using a pool of threads for efficiency (starting a new thread comes with some overhead cost that can be avoided through pooling and reuse). `THREAD_POOL_EXECUTOR` is an instance of the JDK class `ThreadPoolExecutor`, which uses a pool of threads that grows and shrinks with demand. In the case of `AsyncTask`, the pool is configured to maintain at least five threads, and expands up to 128 threads.

To execute `AsyncTask` using a specific executor, we invoke the `executeOnExecutor` method, supplying a reference to the executor we want to use, for example:

```
task.executeOnExecutor(AsyncTask.THREAD_POOL_EXECUTOR, params);
```

As the default behavior of `execute` since API level 11 is to run AsyncTasks serially on a single background thread, the following two statements are equivalent:

```
task.execute(params);
task.executeOnExecutor(AsyncTask.SERIAL_EXECUTOR, params);
```

Besides the default executors provided by `AsyncTask`, we can choose to create our own. For example, we might want to allow some concurrency by operating off a small pool of threads, and allow many tasks to be queued if all threads are currently busy.

This is easily achieved by configuring our own instance of `ThreadPoolExecutor` as a static member of one of our own classes—for example, our `Activity` class. Here's how we might configure an executor with a pool of four to eight threads and an effectively infinite queue:

```
private static final Queue<Runnable> QUEUE =
  new LinkedBlockingQueue<Runnable>();
public static final Executor MY_EXECUTOR =
  new ThreadPoolExecutor(4, 8, 1, TimeUnit.MINUTES, QUEUE);
```

The parameters to the constructor indicate the core pool size (4), the maximum pool size (8), the time for which idle additional threads may live in the pool before being removed (1), the unit of time (`minutes`), and the queue to use when the pool threads are busy.

Using our own `Executor` is then as simple as invoking our `AsyncTask` as follows:

```
task.executeOnExecutor(MY_EXECUTOR, params);
```

Common AsyncTask issues

As with any powerful programming abstraction, `AsyncTask` is not entirely free from issues and compromises.

Fragmentation issues

In the *Controlling the level of concurrency* section, we saw how `AsyncTask` has evolved with new releases of the Android platform, resulting in behavior that varies with the platform of the device running the task, which is a part of the wider issue of fragmentation.

The simple fact is that if we target a broad range of API levels, the execution characteristics of our AsyncTasks—and therefore, the behavior of our apps— can vary considerably on different devices. So what can we do to reduce the likelihood of encountering `AsyncTask` issues due to fragmentation?

The most obvious approach is to deliberately target devices running at least Honeycomb, by setting a `minSdkVersion` of 11 in the **Android Manifest** file. This neatly puts us in the category of devices, which, by default, execute AsyncTasks serially, and therefore, much more predictably.

However, this significantly reduces the market reach of our apps. At the time of writing in September 2013, more than 34 percent of Android devices in the wild run a version of Android in the danger zone between API levels 4 and 10.

A second option is to design our code carefully and test exhaustively on a range of devices—always commendable practices of course, but as we've seen, concurrent programming is hard enough without the added complexity of fragmentation, and invariably, subtle bugs will remain.

A third solution that has been suggested by the Android development community is to reimplement `AsyncTask` in a package within your own project, then extend your own `AsyncTask` class instead of the SDK version. In this way, you are no longer at the mercy of the user's device platform, and can regain control of your AsyncTasks. Since the source code for `AsyncTask` is readily available, this is not difficult to do.

Activity lifecycle issues

Having deliberately moved any long-running tasks off the main thread, we've made our applications nice and responsive—the main thread is free to respond very quickly to any user interaction.

Unfortunately, we have also created a potential problem for ourselves, because the main thread is able to finish the Activity before our background tasks complete. Activity might finish for many reasons, including configuration changes caused the by the user rotating the device (the default behavior of `Activity` on a change in orientation is to restart with an entirely new instance of the activity).

If we continue processing a background task after the Activity has finished, we are probably doing unnecessary work, and therefore wasting CPU and other resources (including battery life), which could be put to better use.

Also, any object references held by the AsyncTask will not be eligible for garbage collection until the task explicitly nulls those references or completes and is itself eligible for **GC (garbage collection)**. Since our AsyncTask probably references the Activity or parts of the View hierarchy, we can easily leak a significant amount of memory in this way.

A common usage of AsyncTask is to declare it as an anonymous inner class of the host Activity, which creates an implicit reference to the Activity and an even bigger memory leak.

There are two approaches for preventing these resource wastage problems.

Handling lifecycle issues with early cancellation

First, and most obviously, we can synchronize our AsyncTask lifecycle with that of the Activity by canceling running tasks when our Activity is finishing.

When an Activity finishes, its lifecycle callback methods are invoked on the main thread. We can check to see why the lifecycle method is being called, and if the Activity is finishing, cancel the background tasks. The most appropriate Activity lifecycle method for this is onPause, which is guaranteed to be called before the Activity finishes.

```
protected void onPause() {
  super.onPause();
  if ((task != null) && (isFinishing()))
    task.cancel(false);
}
```

If the Activity is not finishing—say because it has started another Activity and is still on the back stack—we might simply allow our background task to continue to completion.

Handling lifecycle issues with retained headless fragments

If the Activity is finishing because of a configuration change, it may still be useful to complete the background task and display the results in the restarted Activity. One pattern for achieving this is through the use of retained Fragments.

Fragments were introduced to Android at API level 11, but are available through a support library to applications targeting earlier API levels. All of the downloadable examples use the support library, and target API levels 7 through 19. To use Fragments, our Activity must extend the FragmentActivity class.

The Fragment lifecycle is closely bound to that of the host Activity, and a fragment will normally be disposed when the activity restarts. However, we can explicitly prevent this by invoking setRetainInstance(true) on our Fragment so that it survives across Activity restarts.

Typically, a Fragment will be responsible for creating and managing at least a portion of the user interface of an Activity, but this is not mandatory. A Fragment that does not manage a view of its own is known as a **headless** Fragment.

Isolating our AsyncTask in a retained headless Fragment makes it less likely that we will accidentally leak references to objects such as the View hierarchy, because the AsyncTask will no longer directly interact with the user interface. To demonstrate this, we'll start by defining an interface that our Activity will implement:

```
public interface AsyncListener<Progress, Result> {
  void onPreExecute();
  void onProgressUpdate(Progress... progress);
  void onPostExecute(Result result);
  void onCancelled(Result result);
}
```

Next, we'll create a retained headless Fragment, which wraps our AsyncTask. For brevity, doInBackground is omitted, as it is unchanged from the previous examples — see the downloadable samples for the complete code.

```
public class PrimesFragment extends Fragment {
  private AsyncListener<Integer,BigInteger> listener;
  private PrimesTask task;
  public void onCreate(Bundle savedInstanceState) {
    super.onCreate(savedInstanceState);
    setRetainInstance(true);
    task = new PrimesTask();
    task.execute(2000);
  }

  public void onAttach(Activity activity) {
    super.onAttach(activity);
    listener = (AsyncListener<Integer,BigInteger>)activity;
  }

  public void onDetach() {
    super.onDetach();
    listener = null;
  }
```

```
class PrimesTask
extends AsyncTask<Integer, Integer, BigInteger>{
  protected void onPreExecute() {
    if (listener != null) listener.onPreExecute();
    }

  protected void onProgressUpdate(Integer... values) {
    if (listener != null)
      listener.onProgressUpdate(values);
  }

  protected void onPostExecute(BigInteger result) {
    if (listener != null)
      listener.onPostExecute(result);
  }

  protected void onCancelled(BigInteger result) {
    if (listener != null)
      listener.onCancelled(result);
  }

  // … doInBackground elided for brevity …
  }
}
```

We're using the `Fragment` lifecycle methods (`onAttach` and `onDetach`) to add or remove the current `Activity` as a listener, and `PrimesTask` delegates directly to it from all of its main-thread callbacks.

Now, all we need is the host `Activity` that implements `AsyncListener` and uses `PrimesFragment` to implement its long-running task. The full source code is available to download from the Packt Publishing website, so we'll just take a look at the highlights.

First, the code in the button's `OnClickListener` now checks to see if `Fragment` already exists, and only creates one if it is missing:

```
FragmentManager fm = getSupportFragmentManager();
PrimesFragment primes =
  (PrimesFragment)fm.findFragmentByTag("primes");
if (primes == null) {
  primes = new PrimesFragment();
  FragmentTransaction transaction = fm.beginTransaction();
  transaction.add(primes, "primes").commit();
}
```

If our `Activity` has been restarted, it will need to re-display the progress dialog when a progress update callback is received, so we check and show it, if necessary, before updating the progress bar:

```
public void onProgressUpdate(Integer... progress) {
  if (dialog == null)
    prepareProgressDialog();
  progress.setProgress(progress[0]);
}
```

Finally, `Activity` will need to implement the `onPostExecute` and `onCancelled` callbacks defined by `AsyncListener`. Both methods will update the `resultView` as in the previous examples, then do a little cleanup — dismissing the dialog and removing `Fragment` as its work is now done:

```
public void onPostExecute(BigInteger result) {
  resultView.setText(result.toString());
  cleanUp();
}

public void onCancelled(BigInteger result) {
  resultView.setText("cancelled at " + result);
  cleanUp();
}

private void cleanUp() {
  dialog.dismiss();
  dialog = null;
  FragmentManager fm = getSupportFragmentManager();
  Fragment primes = fm.findFragmentByTag("primes");
  fm.beginTransaction().remove(primes).commit();
}
```

Applications of AsyncTask

Now that we have seen how to use `AsyncTask`, we might ask ourselves when we should use it.

Good candidate applications for `AsyncTask` tend to be relatively short-lived operations (at most, for a second or two), which pertain directly to a specific `Fragment` or `Activity` and need to update its user interface.

`AsyncTask` is ideal for running short, CPU-intensive tasks, such as number crunching or searching for words in large text strings, moving them off the main thread so that it can remain responsive to input and maintain high frame rates.

Blocking I/O operations such as reading and writing text files, or loading images from local files with `BitmapFactory`, are also good use cases for `AsyncTask`.

Of course, there are use cases for which `AsyncTask` is not ideally suited. For anything that requires more than a second or two, we should weigh the cost of performing this operation repeatedly if the user rotates the device, or switches between apps or activities, or whatever else may be going on that we cannot control.

Taking these things into account, and the rate at which complexity increases as we try to deal with them (for example, retained headless fragments!), `AsyncTask` starts to lose its shine for longer operations.

`AsyncTask` is often used to fetch data from remote web servers, but this can fall foul of the `Activity` lifecycle issues we looked at earlier. End users may be working with a flaky 3G or HSDPA connection, where network latencies and bandwidth can vary widely, and a complete HTTP request-response cycle can easily span many seconds. This is especially important when we are uploading significant amount of data, such as an image, as the available bandwidth is often asymmetric.

While we must perform network I/O off the main thread, `AsyncTask` is not necessarily the ideal option—as we'll see later; there are more appropriate constructs available for offloading this kind of work from the main thread.

Summary

In this chapter, we've taken a detailed look at `AsyncTask` and how to use it to write responsive applications that perform operations without blocking the main thread.

We saw how to keep the user informed of the progress, and even allow them to cancel operations early. We also learned how to deal with issues that can arise when the `Activity` lifecycle conspires against our background tasks.

Finally, we considered when to use `AsyncTask`, and when it might not be appropriate.

In the next chapter, we'll take a look at some lower-level constructs—fundamental building blocks on which the other concurrency mechanisms of the platform, including `AsyncTask`, are built.

3
Distributing Work with Handler and HandlerThread

In *Chapter 2, Staying Responsive with AsyncTask*, we familiarized ourselves with the most well-known concurrency construct of the Android platform. What is perhaps less well known are the mechanics of how `AsyncTask` coordinates work between background threads and the main thread.

In this chapter we'll meet some of the lower-level constructs that `AsyncTask` builds on to get its work done.

We'll see how to defer tasks to happen in the future on the main thread, whether that is as soon as possible, after a specified delay, or at a specified time, and we'll apply the same concepts to scheduling work on background threads and coordinating the results with the main thread.

In this chapter we will cover the following topics:

- Understanding `Looper`
- Building responsive apps with `Handler`
- Scheduling work with `post`
- Canceling a pending `Runnable`
- Scheduling work with `send`
- Canceling pending Messages
- Multithreaded programming with `Handler`
- Building responsive apps with `HandlerThread`
- `Handler` programming issues
- Applications of `Handler` and `HandlerThread`

Understanding Looper

Before we can understand Handler, we need to meet the aptly named Looper. Looper is a simple class that quite literally loops forever, waiting for Messages to be added to its queue and dispatching them to target Handlers. It is an implementation of a common UI programming concept known an Event Loop.

To set up a Looper thread, we need to invoke two static methods of Looper — prepare and loop — from within the thread that will handle the message loop.

```
class SimpleLooper extends Thread {
    public void run() {
        Looper.prepare();
        Looper.loop();
    }
}
```

That was easy; however, the SimpleLooper class defined here provides no way to add messages to its queue, which is where Handler comes in.

Handler serves two purposes — to provide an interface to submit Messages to its Looper queue and to implement the callback for processing those Messages when they are dispatched by the Looper.

To attach a Handler to SimpleLooper, we need to instantiate the Handler from within the same thread that prepared the Looper. The Handler is associated with the main Looper by the super-class constructor, which obtains the Looper using a static method.

```
class SimpleLooper extends Thread {
    public Handler handler;
    public void run() {
        Looper.prepare();
        handler = new Handler();
        Looper.loop();
    }
}
```

Once started, the Looper thread will wait (Object.wait) inside Looper.loop() for Messages to be added to its queue.

When another thread adds a Message to the queue it will notify (Object.notify) the waiting thread, which will then dispatch the Message to its target Handler by invoking the Handler's dispatchMessage method.

The wait/notify mechanism is a very efficient way of suspending activity on a particular thread while there is no work for it to do, so that it doesn't waste CPU cycles, then resuming work once there is something for it to do.

Because `dispatchMessage` is invoked by the thread running the message loop, we can send Messages to the `Handler` from any thread and they will always be dispatched by the message loop thread as shown in the following diagram:

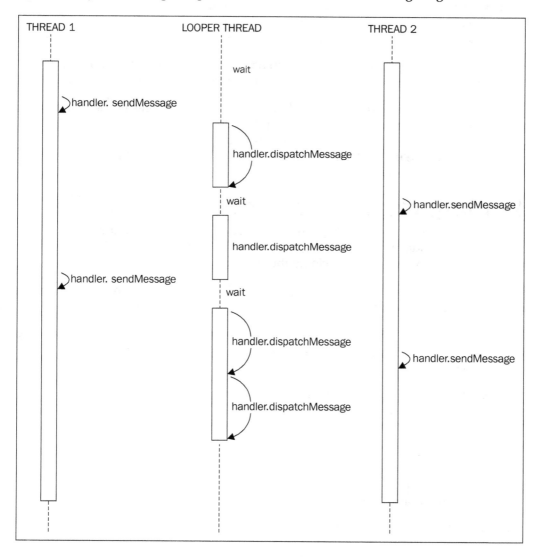

We already saw that we can create our own `Looper` threads, but here's the fun part and this may come as a surprise: the main thread is in fact a `Looper` thread, as we can see in the following stack trace:

```
at example.handler.MyActivity.onCreate(Example1Activity.java:19)
at android.app.Activity.performCreate(Activity.java:5206)
...
at android.os.Handler.dispatchMessage(Handler.java:99)
at android.os.Looper.loop(Looper.java:137)
at android.app.ActivityThread.main(ActivityThread.java:4898)
```

That's right! Everything that happens in an `Activity` lifecycle callback method is running in a `dispatchMessage` call invoked by the main `Looper`!

The interesting thing to realize here is that we can send messages to the main thread from any other thread (or even from the main thread itself) and in doing so, hand over work from background threads to the main thread—for example, to have it update the user interface with the results of background processing.

Building responsive apps with Handler

The `Handler` class is fundamental to the infrastructure of Android apps—together with `Looper`. It underpins everything that the main thread does—including the invocation of the `Activity` lifecycle methods.

While `Looper` takes care of dispatching work on its message-loop thread, `Handler` provides the means to add work to the message queue belonging to a `Looper`.

We can create a `Handler` to submit work to be processed on the main thread simply by creating a new instance of `Handler` from an `Activity` lifecycle method such as `onCreate`, shown as follows:

```
protected void onCreate(Bundle savedInstanceState) {
    super.onCreate(savedInstanceState);
    Handler handler = new Handler();
    // ...
}
```

We could also be explicit that we want to submit work to the main thread by passing a reference to the main `Looper` instance into the `Handler` constructor.

```
Handler handler = new Handler(Looper.getMainLooper());
```

Exactly what we mean by "work" can be described by subclasses of `java.lang.Runnable`, or instances of `android.os.Message`. We can post Runnables to a `Handler` instance or send Messages to it, and it will add them to the queue belonging to the associated `Looper` instance.

Scheduling work with post

We can post work to a `Handler` very simply, for example, by creating an anonymous inner `Runnable`:

```
handler.post(new Runnable(){
    public void run() {
        // do some work on the main thread.
    }
});
```

The `Looper` instance to which the `Handler` is bound works its way through the queue, executing each `Runnable` as soon as possible. Posting with the `post` method simply adds a new `Runnable` at the end of the queue.

If we want our `Runnable` to take priority over anything currently in the queue, we can post it to the front of the queue ahead of existing work:

```
handler.postAtFrontOfQueue(new Runnable(){
    public void run() {
        // do some work on the main thread.
    }
});
```

In a single-threaded application it might seem like there is not a whole lot to be gained from posting work to the main thread like this, but breaking things down into small tasks that can be interleaved and potentially reordered is very useful for maintaining responsiveness.

This becomes clear if we consider the question: what if we want to schedule some work in 10 seconds time? One way to do this would be to `Thread.sleep` the main thread for 10 seconds, but that would mean the main thread can't do anything else in the meantime. Also, remember that we must never block the main thread! The alternative is to post a delayed `Runnable` to the `Handler`.

```
handler.postDelayed(new Runnable(){
    public void run() {
        // do some work on the main thread
        // in 10 seconds time
```

```
    }
}, TimeUnit.SECONDS.toMillis(10));
```

We can still post additional work for execution in the meantime, and our delayed `Runnable` instance will execute after the specified delay. Note that we're using the `TimeUnit` class from the `java.lang.concurrent` package to convert seconds to milliseconds.

A further scheduling option for posted work is `postAtTime`, which schedules `Runnable` to execute at a particular time relative to the system uptime (how long it has been since the system booted):

```
handler.postAtTime(new Runnable(){
    public void run() {
        // … do some work on the main thread
    }
}, SystemClock.uptimeMillis() + TimeUnit.SECONDS.toMillis(10));
```

Since `postDelayed` is implemented in terms of an offset from the `SystemClock` uptime, it is usually easier to just use `postDelayed`.

As we'll see in the *Issues* section, posting an anonymous `Runnable` makes for concise examples but when used with `postDelayed` or `postAtTime` requires care to avoid potential resource leakage.

Since we instantiated our `Handler` on the main thread, all work submitted to it executes on the main thread. This means that we must not submit long running operations to this particular `Handler`, but we can safely interact with the user interface:

```
handler.post(new Runnable(){
    public void run() {
        TextView text = (TextView) findViewById(R.id.text);
        text.setText("updated on the UI thread");
    }
});
```

This applies regardless of which thread posts the `Runnable`, which makes `Handler` an ideal way to send the results of work performed by other threads to the main thread.

```
Thread thread = new Thread(){
    public void run(){
        final BigInteger result = calculatePrime(500);
        handler.post(new Runnable(){
```

```
            public void run() {
                TextView text = (TextView)
                    findViewById(R.id.text);
                text.setText(result.toString());
            }
        });
    }
};
thread.setPriority(Thread.MIN_PRIORITY);
thread.start();
```

 If you start your own threads for the background work, make sure to set the priority to `Thread.MIN_PRIORITY` to avoid starving the main thread of CPU time.

`Handler` is so fundamental that it is integrated right into the `View` class hierarchy, so we can rewrite the last example as follows:

```
final TextView text = (TextView) findViewById(R.id.text);
Thread thread = new Thread() {
    public void run() {
        final BigInteger result = calculatePrime(500);
        text.post(new Runnable() {
            public void run() {
                text.setText(result.toString());
            }
        });
    }
};
thread.setPriority(Thread.MIN_PRIORITY);
thread.start();
```

For comparison, both of the previous examples are roughly equivalent to the following `AsyncTask` code:

```
new AsyncTask<Void, Void, BigInteger>() {
    public BigInteger doInBackground(Void... params) {
        return calculatePrime(500);
    }
    public void onPostExecute(BigInteger result) {
        TextView text = (TextView) findViewById(R.id.text);
        text.setText(result.toString());
    }
}.execute();
```

When writing code in an `Activity` class, there is an alternative way of executing a Runnable on the main thread using the `runOnUiThread` method of `Activity`. If invoked from a background thread, the Runnable will be posted to a Handler instance attached to the main thread. If invoked from the main thread, the Runnable will be executed immediately.

Canceling a pending Runnable

We can cancel a pending operation by removing a posted Runnable callback from the queue.

```
final Runnable runnable = new Runnable(){
    public void run() {
        // … do some work
    }
};
handler.postDelayed(runnable, TimeUnit.SECONDS.toMillis(10));
Button cancel = (Button) findViewById(R.id.cancel);
cancel.setOnClickListener(new OnClickListener(){
    public void onClick(View v) {
        handler.removeCallbacks(runnable);
    }
});
```

Notice that in order to be able to specify what to remove, we must keep a reference to the Runnable instance, and that cancellation applies only to pending tasks—it does not attempt to stop a Runnable that is already mid-execution.

Scheduling work with send

When we post a Runnable, we can—as seen in the previous examples—define the work at the call site with an anonymous Runnable. As such, the Handler does not know in advance what kind of work it might be asked to perform.

If we often need to perform the same work from different call sites we could define a static or top-level Runnable class that we can instantiate from different call sites.

Alternatively, we can turn the approach on its head by sending Messages to a Handler, and defining the Handler to react appropriately to different Messages.

Taking a simple example, let's say we want our `Handler` to display `hello` or goodbye, depending on what type of `Message` it receives. To do that, we'll extend `Handler` and override its `handleMessage` method.

```
public static class SpeakHandler extends Handler {
    public static final int SAY_HELLO = 0;
    public static final int SAY_BYE = 1;
    @Override
    public void handleMessage(Message msg) {
        switch(msg.what) {
            case SAY_HELLO:
                sayWord("hello"); break;
            case SAY_BYE:
                sayWord("goodbye"); break;
            default:
                super.handleMessage(msg);
        }
    }
    private void sayWord(String word) { … }
}
```

Here we've implemented the `handleMessage` method to expect messages with two different `what` values and react accordingly.

 If you look carefully at the Speak Handler class example explained earlier, you'll notice that we defined it as a static class. Subclasses of `Handler` should always be declared as top-level or static classes to avoid inadvertent memory leaks!

To attach an instance of our `Handler` to the main thread, we simply instantiate it from any method which runs on the main thread:

```
private Handler handler;
protected void onCreate(Bundle savedInstanceState) {
    super.onCreate(savedInstanceState);
    handler = new SpeakHandler();
    // …
}
```

Remember that we can send Messages to this `Handler` from any thread, and they will be processed by the main thread. We send Messages to our `Handler` as shown:

```
handler.sendEmptyMessage(SAY_HELLO);
```

Messages may also carry an object payload as context for the execution of the message. Let's extend our example to allow our `Handler` to say any word that the message sender wants:

```
public static class SpeakHandler extends Handler {
    public static final int SAY_HELLO = 0;
    public static final int SAY_BYE = 1;
    public static final int SAY_WORD = 2;
    @Override
    public void handleMessage(Message msg) {
        switch(msg.what) {
            case SAY_HELLO:
                sayWord("hello"); break;
            case SAY_BYE:
                sayWord("goodbye"); break;
            case SAY_WORD:
                sayWord((String)msg.obj); break;
            default:
                super.handleMessage(msg);
        }
    }
    private void sayWord(String word) { … }
}
```

Within our `handleMessage` method, we can access the payload of the `Message` directly by accessing the public `obj` property. The `Message` payload can be set easily via alternative static `obtain` methods.

```
handler.sendMessage(
    Message.obtain(handler, SpeakHandler.SAY_WORD, "tada!"));
```

While it should be quite clear what this code is doing, you might be wondering why we didn't create a new instance of `Message` by invoking its constructor, and instead invoked its static method `obtain`.

The reason is efficiency. Messages are used only briefly—we instantiate, dispatch, handle, and then discard them. So if we create new instances for each we are creating work for the garbage collector.

Garbage collection is expensive, and the Android platform goes out of its way to minimize object allocation whenever it can. While we can instantiate a new `Message` object if we wish, the recommended approach is to obtain one, which re-uses `Message` instances from a pool and cuts down on garbage collection overhead.

Just as we can schedule Runnables with the variants of the post method, we can schedule Messages with variants of send.

```
handler.sendMessageAtFrontOfQueue(msg);
handler.sendMessageAtTime(msg, time);
handler.sendMessageDelayed(msg, delay);
```

There are also empty-message variants for convenience, when we don't have a payload.

```
handler.sendEmptyMessageAtTime(what, time);
handler.sendEmptyMessageDelayed(what, delay);
```

Canceling pending Messages

Canceling sent Messages is also possible, and actually easier than canceling posted Runnables because we don't have to keep a reference to the Messages that we might want to cancel—instead we can just cancel Messages by their what value.

```
handler.removeMessages(SpeakHandler.SAY_WORD);
```

Note that just as with posted Runnables, Message cancellation only removes pending operations from the queue—it does not attempt to stop an operation which is already being executed.

Composition versus Inheritance

So far we subclassed Handler to override its handleMessage method, but that isn't our only option. We can prefer composition over inheritance by passing an instance of Handler.Callback during Handler construction.

```
public static class Speaker implements Handler.Callback {
    public static final int SAY_HELLO = 0;
    public static final int SAY_BYE = 1;
    public static final int SAY_WORD = 2;
    @Override
    public boolean handleMessage(Message msg) {
        switch(msg.what) {
            case SAY_HELLO:
                sayWord("hello"); break;
            case SAY_BYE:
                sayWord("goodbye"); break;
            case SAY_WORD:
                sayWord((String)msg.obj); break;
            default:
```

```
                return false;
            }
            return true;
        }
        private void sayWord(String word) { … }
    }
```

Notice that the signature of handleMessage is slightly different here — we must return a boolean indicating whether or not the Message was handled. To create a Handler that uses our Callback, simply pass the Callback during Handler construction.

```
Handler handler = new Handler(new Speaker());
```

If we return false from the handleMessage method of our Callback, the Handler will invoke its own handleMessage method, so we could choose to use a combination of inheritance and composition to implement default behavior in a Handler subclass, and mix in special behavior by passing in an instance of Handler.Callback.

```
public static class SpeakHandler extends Handler {
    public static final int SAY_HELLO = 0;
    public static final int SAY_BYE = 1;
    public SpeakHandler(Callback callback) {
        super(callback);
    }
    @Override
    public void handleMessage(Message msg) {
        switch(msg.what) {
            case SAY_HELLO:
                sayWord("hello"); break;
            case SAY_BYE:
                sayWord("goodbye"); break;
            default:
                super.handleMessage(msg);
        }
    }
    private void sayWord(String word) { … }
}
public static class Speaker implements Handler.Callback {
    public static final int SAY_WORD = 2;
    @Override
    public boolean handleMessage(Message msg) {
        if (msg.what == SAY_WORD) {
```

```
                sayWord((String)msg.obj);
                return true;
            }
            return false;
        }
        private void sayWord(String word) { … }
    }
    Handler h = new SpeakHandler(new Speaker());
    h.sendMessage(Message.obtain(handler, Speaker.SAY_WORD, "!?"));
```

With `SpeakHandler` set up like this, we can easily send Messages from any thread to update the user interface. Sending from a background thread or the main thread itself is exactly the same—just obtain a `Message` and send it via `SpeakHandler`.

Multithreaded example

Let's extend our example to bind the app to a network socket and echo lines of text it receives from the socket to the screen. Listening for lines of text from the network socket is a blocking operation, so we must not do it from the main thread.

We'll start a background thread, then bind the server socket and wait for a client to connect over the network. When this background thread receives text from the socket, it will send it in a `Message` to the `SpeakHandler` instance, which will update the user interface on the main thread.

```
    static class Parrot extends Thread {
        private Handler handler;
        private InetAddress address;
        private ServerSocket server;

        public Parrot(InetAddress address, Handler handler) {
            this.handler = handler;
            this.address = address;
            setPriority(Thread.MIN_PRIORITY);
        }

        public void run() {
            try {
                server = new ServerSocket(4444, 1, address);
                while (true) {
                    Socket client = server.accept();
                    handler.sendMessage(Message.obtain(handler,
                        SpeakHandler.SAY_HELLO));
                    BufferedReader in = new BufferedReader(
```

```
                    new InputStreamReader(
                        client.getInputStream()));
                String word;
                while (!"bye".equals(word = in.readLine())) {
                    handler.sendMessage(Message.obtain(handler,
                        SpeakHandler.SAY_WORD, word));
                }
                client.close();
                handler.sendMessage(
                    Message.obtain(
                        handler, SpeakHandler.SAY_BYE));
            }
        } catch (Exception exc) {
            Log.e(TAG, exc.getMessage(), exc);
        }
    }
}
```

Let's have a quick look at the key elements of the run method. First, we bind a socket to port 4444 (any port higher than 1024 will do) on the given address, which allows us to listen for incoming connections. The second parameter says we're only allowing one connection at a time:

```
server = new ServerSocket(4444, 1, address);
```

We loop forever, or until an exception is thrown. Inside the loop we wait for a client to connect:

```
Socket client = server.accept();
```

The accept method blocks, so our background thread is suspended until a client makes a connection. As soon as a client connects, we send a SAY_HELLO message to our Handler, so the main thread will update the user interface for us:

```
handler.sendMessage(
    Message.obtain(handler, SpeakHandler.SAY_HELLO));
```

Next, we wrap a buffering reader around our socket's input stream, and loop on its readLine method, which blocks until a line of text is available.

```
BufferedReader in = new BufferedReader(
    new InputStreamReader(client.getInputStream()));
String word;
while (!"bye".equals(word = in.readLine())) {
```

When we receive some text from the socket, we send it in a `Message` to our `Handler`.

```
handler.sendMessage(
    Message.obtain(handler, SpeakHandler.SAY_WORD, word));
```

If we receive `bye`, we'll break out of the loop, disconnect this client, and say goodbye:

```
client.close();
handler.sendMessage(
    Message.obtain(handler, SpeakHandler.SAY_BYE));
```

To obtain the address of the device, we can use the Wi-Fi service. We'll need to convert the value we get from the Wi-Fi service to an instance of the `InetAddress` class, elided here for brevity.

```
private InetAddress getAddress() {
    WifiManager wm = (WifiManager)getSystemService(WIFI_SERVICE);
    return asInetAddress(wm.getConnectionInfo().getIpAddress());
}
```

Binding the socket and using the Wi-Fi service requires permissions to be requested in the Android manifest:

```
<uses-permission android:name="android.permission.INTERNET"/>
<uses-permission
    android:name="android.permission.ACCESS_WIFI_STATE"/>
```

Finally we need to create and start `Parrot` in a suitable `Activity` lifecycle method, for example, `onResume`. To make it easy to connect, we'll display the address and port on the screen.

```
TextView view = (TextView) findViewById(R.id.speak);
if (parrot == null) {
    InetAddress address = getAddress();
    parrot = new Parrot(address, handler);
    parrot.start();
    view.setText("telnet " + address.getHostAddress() + " 4444");
}
```

We should also remove pending messages in `onPause`, and disconnect the socket if the `Activity` is finishing — the complete code is available for download from the accompanying website.

When the app starts, you'll see a message on the device screen like **telnet 192.168.0.4 4444**. To connect and send messages to your device's screen, open the command prompt on your development computer and copy the text from your device screen to the command prompt.

You should see the following output:

```
Trying 192.168.0.4...
Connected to 192.168.0.4.
Escape character is '^]'.
```

Congratulations! You're connected. Enter some words in the command prompt and they'll appear on your device screen. Enter bye to disconnect.

Sending Messages versus posting Runnables

It is worth spending a few moments to consider the difference between posting Runnables and sending Messages.

The runtime difference mostly comes down to efficiency. Creating new instances of Runnable each time we want our Handler to do something adds garbage collection overhead, while sending messages re-uses Message instances, which are sourced from an application-wide pool.

The difference at development time is between allowing code at the call site to specify arbitrary work for the Handler to do (Runnable.run) potentially spreading similar code throughout the codebase, versus the Handler defining the work it is prepared to do in a single place (Handler.handleMessage).

For prototyping and small one-offs, posting Runnables is quick and easy, while the advantages of sending Messages tend to grow with the size of the application. It should be said that Message sending is more "The Android Way", and is used throughout the platform to keep garbage to a minimum and apps running smoothly.

Building responsive apps with HandlerThread

So far we only really considered Handler as a way to request work be performed on the main thread – we've submitted work from the main thread to itself, and from a background thread to the main thread.

In fact we can bind Handlers to any thread we create, and in doing so allow any thread to submit work for another thread to execute. For example, we can submit work from the main thread to a background thread or from one background thread to another.

We saw one way of setting up a Looper thread with a Handler in the *Understanding Looper* section earlier in this chapter, but there's an easier way using a class provided by the SDK for exactly this purpose, android.os.HandlerThread.

When we create a `HandlerThread`, we specify two things: a name for the thread, which can be helpful when debugging; and its priority, which must be selected from the set of static values in the `android.os.Process` class.

```
HandlerThread thread =
    new HandlerThread("bg", Process.THREAD_PRIORITY_BACKGROUND);
```

Thread priorities in Android are mapped onto Linux "nice" levels, which govern how often a thread gets to run.

In addition to prioritization, Android limits CPU resources using Linux cgroups. Threads that are of background priority are moved into the `bg_non_interactive` cgroup, which is limited to 5 percent of available CPU if threads in other groups are busy.

Adding `THREAD_PRIORITY_MORE_FAVORABLE` to `THREAD_PRIORITY_BACKGROUND` when configuring your `HandlerThread` moves the thread into the default cgroup, but always consider whether it is really necessary—it often isn't!

`HandlerThread` extends `java.lang.Thread`, and we must `start()` it before it will actually begin processing its queue:

```
thread.start();
```

Now we need a `Handler` through which we can pass work to our `HandlerThread`. So we create a new instance, but this time we parameterize the constructor with the `Looper` associated with our `HandlerThread`.

```
Handler handler = new Handler(thread.getLooper());
```

That's all there is to it—we can now post Runnables for our `HandlerThread` to execute:

```
handler.post(new Runnable(){
    public void run() {
        // ... do some work in the background thread
    }
});
```

To send Messages to our `HandlerThread`, we'll need to override `handleMessage` or provide a `Callback` via the alternate `Handler` constructor:

```
Handler.Callback callback = new Handler.Callback(){ … };
Handler handler = new Handler(thread.getLooper(), callback);
```

If we create a HandlerThread to do background work for a specific Activity, we will want to tie the HandlerThread's lifecycle closely to that of the Activity to prevent resource leaks.

A HandlerThread can be shut down by invoking quit, which will stop the HandlerThread from processing any more work from its queue. A quitSafely method was added at API level 18, which causes the HandlerThread to process all remaining tasks before shutting down. Once a HandlerThread has been told to shut down, it will not accept any further tasks.

Just as we did in the previous chapter with AsyncTask, we can use the Activity lifecycle methods to determine when we should quit the HandlerThread.

```
private HandlerThread thread;
protected void onCreate(Bundle savedInstanceState) {
    super.onCreate(savedInstanceState);
    thread = new HandlerThread( … );
}
protected void onPause() {
    super.onPause();
    if ((thread != null) && (isFinishing()))
        thread.quit();
}
```

Handler programming issues

The Handler class is truly fundamental to the Android platform and is used widely throughout, but there are plenty of ways we can get ourselves into trouble if we aren't careful.

Leaking implicit references

The biggest worry when using Handler within an Activity is resource leakage which, just as with AsyncTask, is very easy to do. Here's one of our earlier examples:

```
final Runnable runnable = new Runnable(){
    public void run() {
        // … do some work
    }
};
handler.postDelayed(runnable, TimeUnit.SECONDS.toMillis(10));
```

By declaring an anonymous inner `Runnable` inside an activity, we have made an implicit reference to that containing `Activity` instance. We've then posted the `Runnable` to a handler and told it to execute in 10 seconds time.

If the `Activity` finishes before the 10 seconds are up, it cannot yet be garbage collected because the implicit reference in our `Runnable` means that the `Activity` is still reachable by live objects.

So, although it makes for a concise example, it is not a good idea in practice to post non-static Runnables onto the main thread's Handler queue (especially with `postDelayed` or `postAtTime`) unless we're very careful to clean up after ourselves.

One way to minimize this problem is to avoid using non-static inner classes; for example, by always declaring Runnables as top-level classes in their own file, or as `static` classes in an `Activity` subclass. This means that references must be explicit, which makes them easier to spot and nullify.

In addition, we can cancel pending tasks during `Activity` lifecycle callbacks such as `onPause`. This is easiest if we're working with Messages since we can remove them by their `what` value, and don't have to keep references as we would with Runnables.

For `HandlerThread` instances we've created, we should make sure to `quit` when the `Activity` is finishing, which will prevent further execution and free up the `Runnable` and `Message` objects for garbage collection.

Leaking explicit references

If we are to interact with the user interface, we'll at least need a reference to an object in the `View` hierarchy, which we might pass into our static or top-level Runnable's constructor.

```
static class MyRunnable implements Runnable {
    private View view;
    public MyRunnable(View view) {
        this.view = view;
    }
    public void run() {
        // … do something with the view.
    }
}
```

However, by keeping a strong reference to the `View`, we are again subject to potential memory leaks if our `Runnable` outlives the `View`; for example, if some other part of our code removes this `View` from the display before our `Runnable` executes.

One solution to this is to use a weak reference, and check for null before using the referenced View.

```
static class MyRunnable implements Runnable {
    private WeakReference<View> view;
    public MyRunnable(View view) {
        this.view = new WeakReference<View>(view);
    }
    public void run() {
        View v = view.get(); // might return null
        if (v != null) {
            // … do something with the view.
        }
    }
}
```

If you haven't used WeakReference before, what it gives us is a way to refer to an object only for as long as some other live object has a stronger reference to it (for example, a "normal" property reference).

When all strong references are garbage collected, our WeakReference will also lose its reference to the View, get() will return null, and the View will be garbage collected.

This fixes the resource leakage problem, but we must always check for null before using the returned object, to avoid potential NullPointerException's.

If we're sending Messages to our Handler and expecting it to update the user interface, it will also need a reference to the View hierarchy. A nice way to manage this is to attach and detach the Handler from onResume and onPause.

```
private static class MyHandler extends Handler {
    private TextView view;
    public void attach(TextView view) {
        this.view = view;
    }
    public void detach() {
        view = null;
    }
    @Override
    public void handleMessage(Message msg) {
        //…
    }
}
```

Applications of Handler and HandlerThread

The `Handler` class is incredibly versatile, which makes its range of applications very broad.

So far we looked at `Handler` and `HandlerThread` in the context of the `Activity` lifecycle, which constrains the sort of applications that make sense—ideally we do not want to perform long-running operations (more than a second or so) at all in this context.

With that constraint in mind, good candidate uses include performing calculations, string processing, reading and writing small files on the filesystem, and reading or writing to local databases using a background `HandlerThread`.

We might consider `AsyncTask` instead for one-off's, or if we want to display progress or cancel a task part way through. The Android platform also uses `Handler` extensively as a mechanism for abstracting the work that needs doing from the thread that will do it. A nice example of this can be found in `android.hardware.SensorManager`, which allows listeners to be registered along with a `Handler` so that we can easily handle sensor data in a separate `HandlerThread`.

In the download section available from the Packt Publishing website you'll find an example of processing sensor data in the background, using the magnetic field sensor and the accelerometer in combination to create a simple compass.

Summary

In this chapter we used `Handler` to queue work for the main thread to process, as a means of maintaining responsiveness in a single-threaded application.

We saw the different ways we can define work with `Handler`—arbitrary work defined at the call site with `Runnable`, or predefined work implemented in the `Handler` itself and triggered by message sending.

We learned how to use `Handler` in a multithreaded application to pass work and results back and forth between cooperating threads, performing blocking operations on an ordinary background thread, and communicating the results back to the main thread to update the user interface.

We also met `HandlerThread` and used it to create a background thread with its own `Looper`, allowing us to use these same techniques to queue work for background processing.

This isn't the last we'll see of `Handler` and `HandlerThread`—they can also be usefully put to work in other contexts, as we'll discover in *Chapter 5, Queuing Work with IntentService* and *Chapter 6, Long-running Tasks with Service.*

In the next chapter we'll learn about Loaders—a construct designed to ease asynchronous loading of data on the Android platform.

Asynchronous I/O with Loader

4

The concurrency constructs we've encountered so far have been quite general in purpose, but in this chapter we'll take a look at a construct with a more specific focus — Loader.

In this chapter we will cover the following topics:

- Introducing Loaders
- Building responsive apps with AsyncTaskLoader
- Building responsive apps with CursorLoader
- Combining Loaders
- Applications of Loaders

Introducing Loaders

As the name suggests, the job of Loader is to load data on behalf of other parts of the application, and to make that data available across activities and fragments within the same process.

Loaders were introduced to the Android platform at API level 11, but are available for backwards compatibility through the support libraries. The examples in this chapter use the support library to target API levels 7 through 19.

Provided we implement our Loaders correctly, we get a number of benefits:

- The heavy lifting is automatically performed on a background thread, and the results are safely introduced to the main thread on completion.
- Loaded data can be cached and redelivered on repeat calls for speed and efficiency.

- The framework gives us control over when a Loader instance is destroyed, and allows Loaders to live outside the Activity lifecycle, making their data available across the application and across Activity restarts.

- Loaders monitor their underlying datasource, and reload their data in the background when necessary. The framework includes lifecycle callbacks that allow us to properly dispose of any expensive resources held by our Loaders.

When we use Loaders, we will not do so in isolation, because they form part of a small framework. Loaders are managed objects, and are looked after by a LoaderManager, which takes care of coordinating Loader lifecycle events with the Fragment and Activity lifecycles, and makes the Loader instances available to client code throughout an application.

To connect a Loader with a recipient for the data it loads, the framework provides the LoaderCallbacks interface. LoaderCallbacks requires an implementing class to provide three methods:

```
CursorLoader onCreateLoader(int id, Bundle bundle);
void onLoadFinished(Loader<Cursor> loader, Cursor media);
void onLoaderReset(Loader<Cursor> loader);
```

The onCreateLoader method allows us to instantiate the particular Loader implementation we want. The onLoaderFinished method provides a way to receive the result of background loading in the main thread. The onLoaderReset method gives us a place to perform any cleanup that is needed when Loader is being discarded.

Loader is an abstract class, and does not itself implement any asynchronous behavior. Although we can extend Loader directly, it is more common to use one of the two provided subclasses, AsyncTaskLoader or CursorLoader, depending on our requirements.

AsyncTaskLoader is a general-purpose Loader, which we can subclass when we want to load just about any kind of data from just about any kind of source, and do so off the main thread.

CursorLoader extends AsyncTaskLoader, specializing it to efficiently source data from a local database and manage the associated database Cursor correctly.

Let's begin by implementing a simple AsyncTaskLoader to load a bitmap in the background from the MediaStore.

Building responsive apps with AsyncTaskLoader

`AsyncTaskLoader` is a `Loader` implementation that uses `AsyncTasks` to perform its background work, though this is largely hidden from us when we implement our own subclasses.

We don't need to trouble ourselves about the AsyncTasks—they are completely hidden by `AsyncTaskLoader`—but with what we learned earlier about `AsyncTask`, it is interesting to note that tasks are executed using `AsyncTask.THREAD_POOL_EXECUTOR` to ensure a high degree of concurrency when multiple Loaders are in use.

`Loader` is generically typed so, when we implement it, we need to specify the type of object that it will load—in our case `android.graphics.Bitmap`:

```
public class ThumbnailLoader extends AsyncTaskLoader<Bitmap> {
    // ...
}
```

The `Loader` abstract class requires a Context passed to its constructor, so we must pass a Context up the chain. We'll also need to know which thumbnail to load, so we'll also pass an identifier, `mediaId`:

```
private Integer mediaId;
public ThumbnailLoader(Context context, Integer mediaId){
    super(context);
    this.mediaId = mediaId;
}
```

We don't need to keep our own reference to the Context object—`Loader` exposes a `getContext` method, which we can call from anywhere in our class, where we might need a `Context`.

We can safely pass a reference to an Activity instance as the `Context` parameter, but we should not expect `getContext()` to return the same object!

Loaders potentially live much longer than a single `Activity`, so the `Loader` superclass only keeps a reference to the ApplicationContext to prevent memory leaks.

There are several methods we will need to override, which we'll work through one at a time. The most important is `loadInBackground`—the workhorse of our `Loader`, and the only method which does not run on the main thread:

```
@Override
public Bitmap loadInBackground() {
    //...
}
```

We're going to fetch a thumbnail `Bitmap` from `MediaStore`, which is fairly straightforward but is a potentially slow operation that is good to perform off the main thread.

The following diagram displays the `Loader` Lifecycle, showing callbacks invoked by `Loader` and a typical `AsyncTaskLoader` implementation:

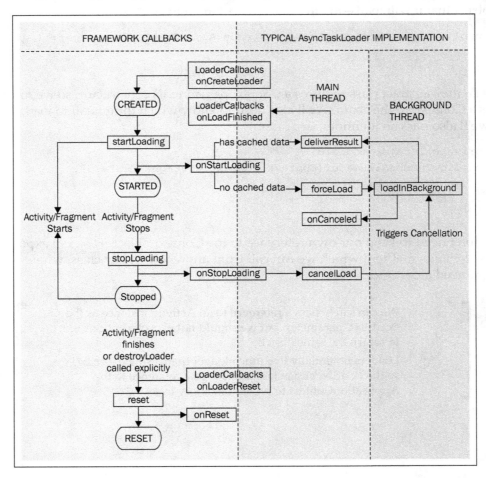

Loading a thumbnail entails some small delay due to blocking I/O reading from permanent storage, but it is also possible that the MediaStore does not yet have a thumbnail of the image we're requesting, in which case it must first read the original image and create a scaled thumbnail from it—a much more expensive operation that we definitely don't want to perform on the main thread!

Luckily we don't need to concern ourselves much with what's going on inside MediaStore—provided we call it on a background thread, we can afford to wait for it to do its thing:

```
@Override
public Bitmap loadInBackground() {
    ContentResolver res = getContext().getContentResolver();
    if (mediaId != null) {
        return MediaStore.Images.Thumbnails.getThumbnail(
            res, mediaId,
            MediaStore.Images.Thumbnails.MINI_KIND, null);
    }
    return null;
}
```

We'll want to cache a reference to the Bitmap object that we're delivering, so that any future calls can just return the same Bitmap immediately. We'll do this by overriding deliverResult:

```
@Override
public void deliverResult(Bitmap data) {
    this.data = data;
    super.deliverResult(data);
}
```

To make our Loader actually work, we still need to override a handful of lifecycle methods that are defined by the Loader base class. First and foremost is onStartLoading:

```
@Override
protected void onStartLoading() {
    if (data != null) {
        deliverResult(data);
    } else {
        forceLoad();
    }
}
```

Here we check our cache (`data`) to see if we have a previously loaded result that we can deliver immediately via `deliverResult`. If not, we trigger a background load to occur — we must do this, or our Loader won't ever load anything. All of this takes place on the main thread.

We now have a minimal working `Loader` implementation, but there is some housekeeping required if we want our `Loader` to play well with the framework.

First of all we need to make sure to clean up the Bitmap when our `Loader` is discarded. `Loader` provides a callback intended for that exact purpose — `onReset`.

```
@Override
protected void onReset() {
    if (data != null)
        data.recycle();
}
```

The framework will ensure that `onReset` is called when `Loader` is being discarded, which will happen when the app exits or when the `Loader` instance is explicitly discarded via `LoaderManager`.

There are two more lifecycle methods, which are important to implement correctly if we want our app to be as responsive as possible: `onStopLoading` and `onCanceled` (be careful of the spelling of `onCanceled` here versus `onCancelled` in most places).

The framework will tell us when it doesn't want us to waste cycles loading data by invoking the `onStopLoading` callback. It may still need the data we have already loaded though, and it may tell us to start loading again, so we should not clean up resources yet. In `AsyncTaskLoader` we'll want to cancel the background work if possible, so we'll just call the superclass `cancelLoad` method:

```
@Override
protected void onStopLoading() {
    cancelLoad();
}
```

Finally, we need to implement `onCanceled` to clean up any data that might be loaded in the background after a cancellation has been issued:

```
@Override
public void onCanceled(Bitmap data) {
    if (data != null)
        data.recycle();
}
```

Depending on the kind of data `Loader` produces, we may not need to worry about cleaning up the result of canceled work—ordinary Java objects will be cleaned up by the garbage collector when they are no longer referenced.

`Bitmap`, however, can be a tricky beast—older versions of Android store the bitmap data in native memory, outside of the normal object heap, requiring a call to `recycle` to avoid potential OutOfMemoryErrors.

So far so good—we have a `Loader`. Now we need to connect it to a client `Activity` or `Fragment`. We know that our `Activity` is going to load an image from an ID reference, and that it will need somewhere to display that image. Let's get these easy bits out of the way first:

```
public class ThumbnailActivity extends Activity {
    private Integer mediaId;
    private ImageView thumb;

    @Override
    protected void onCreate(Bundle savedInstanceState) {
        super.onCreate(savedInstanceState);
        setContentView(R.layout.thumb);
        thumb = (ImageView) findViewById(R.id.thumb);
        mediaId = getMediaIdFromIntent(getIntent());
    }
}
```

To launch this Activity, we will select an image in the gallery app and "send" it to our `Activity`. The `getMediaIdFromIntent` method will extract the ID from the `Intent` we receive:

```
private Integer getMediaIdFromIntent(Intent intent) {
    if (Intent.ACTION_SEND.equals(intent.getAction())) {
        Uri uri = intent.getParcelableExtra(Intent.EXTRA_STREAM);
        return new Integer(uri.getLastPathSegment());
    } else {
        return null;
    }
}
```

In order to make our `Activity` available in the **share via** app selection, we need to specify that it can handle images in an `ACTION_SEND` intent by adding the appropriate intent filter in our `AndroidManifest.xml` file:

```
<activity
    android:name=".example1.ThumbnailActivity"
```

```
            android:label="@string/thumb_activity">
        <intent-filter>
            <action android:name="android.intent.action.SEND"/>
            <data android:mimeType="image/*" />
            ...
        </intent-filter>
    </activity>
```

OK, we're ready to start fitting out our `Activity` to work with `Loader`! First, we need to obtain a reference to `LoaderManager` and call its `initLoader` method.

```
public abstract <D> Loader<D> initLoader(
    int id, Bundle args,
    LoaderManager.LoaderCallbacks<D> callback);
```

The parameters to `initLoader` specify an `int` identifier for `ThumbnailLoader`. We can use this identifier to look up `ThumbnaiLoader` from `LoaderManager` whenever we need to interact with it. Remember that Loaders are not closely bound to the `Activity` lifecycle, so we can use this ID to retrieve the same `Loader` instance in a different `Activity`!

It's generally a good idea to use a public static `int` field to record the ID for each `Loader`, making it easy to use the correct ID from anywhere. A nice way to avoid accidentally re-using the same ID for different Loaders is to use the `hashCode` of a `String` name.

```
public static final int LOADER_ID = "thumb_loader".hashCode();
```

We can also pass a `Bundle` of values—the second parameter—to configure the `Loader` should we need to. For now we'll keep our example simple and just pass `null`.

The third parameter is an object that implements `LoaderCallbacks`. It's quite common to implement the callbacks directly in the `Activity` and pass this when initializing the `Loader` from within the `Activity`:

```
public class ThumbnailActivity extends Activity
implements LoaderManager.LoaderCallbacks<Bitmap> {
    public static final int LOADER_ID =
        "thumb_loader".hashCode();
    private Integer mediaId;
    private ImageView thumb;

    @Override
    protected void onCreate(Bundle savedInstanceState) {
        super.onCreate(savedInstanceState);
        setContentView(R.layout.thumb);
```

```
        thumb = (ImageView) findViewById(R.id.thumb);
        mediaId = getMediaIdFromIntent(getIntent());
        if (mediaId != null)
            getLoaderManager().initLoader(
                LOADER_ID, null, this);
    }

    @Override
    public Loader<Bitmap> onCreateLoader(
        int id, Bundle bundle){
        return null; // todo
    }

    @Override
    public void onLoadFinished(
        Loader<Bitmap> loader, Bitmap bitmap){
        // todo
    }

    @Override
    public void onLoaderReset(Loader<Bitmap> loader) {
        // todo
    }
}
```

When we initialize ThumbnaiLoader via LoaderManager's initLoader method, it will either return an existing Loader with the given ID (LOADER_ID) or, if it doesn't yet have a Loader with that ID, it will invoke the first of the LoaderCallbacks methods—onCreateLoader. This is where we get to choose which type of Loader to instantiate:

```
@Override
public Loader<Bitmap> onCreateLoader(
    int loaderId, Bundle bundle) {
    return new ThumbnailLoader(
        getApplicationContext(), mediaId);
}
```

Notice that the parameters to onCreateLoader receive the same ID and bundle values we passed to the initLoader call.

The implementation of onLoadFinished must take the loaded Bitmap and insert it into our ImageView:

```
@Override
public void onLoadFinished(Loader<Bitmap> loader, Bitmap bitmap) {
    thumb.setImageBitmap(bitmap);
}
```

We're almost done—all that remains is to do any necessary clean up in
onLoaderReset:

```
@Override
protected void onReset() {
    Log.i(LaunchActivity.TAG, getId() + " onReset");

    // if we have a bitmap, make sure it is
    // recycled asap.
    if (data != null) {
        data.recycle();

        data = null;
    }
}
```

Phew! We've loaded an image in the background and displayed it when loading is
completed. When compared with AsyncTask, things here are more complicated—
we've had to write more code and deal with more classes, but the payoff is that
the data is cached for use across Activity restarts and can be used from other
Fragments or Activities.

In the next section we'll implement CursorLoader to efficiently retrieve and display
a list of all images on the device.

Building responsive apps with CursorLoader

CursorLoader is a specialized subclass of AsyncTaskLoader that uses its lifecycle
methods to correctly manage the resources associated with a database Cursor.

A database Cursor is a little like an Iterator, in that it allows you to scroll through
a dataset without having to worry where exactly the dataset is coming from or what
data structure it is a part of.

We're going to use CursorLoader to query the MediaStore for a list of all images on
the device. Because CursorLoader is already implemented to correctly handle all of
the details of working with a Cursor, we don't need to subclass it. We can simply
instantiate it, passing in the information it needs in order to open the Cursor it
should manage for us. We can do this in the onCreateLoader callback:

```
@Override
public CursorLoader onCreateLoader(int id, Bundle bundle) {
    return new CursorLoader(this,
```

```
    MediaStore.Images.Media.EXTERNAL_CONTENT_URI,
    new String[]{
        MediaStore.Images.Media._ID,
        MediaStore.Images.Media.DISPLAY_NAME
    }, "", null, null);
}
```

Just as with the previous example, we'll implement the callbacks in our `Activity` subclass. We're going to use `GridView` to display our loaded results, so we'll implement an `Adapter` interface to supply `views` for its cells, and we'll connect the `Adapter` to the `Cursor` created by our `Loader`:

```
public class MediaStoreActivity extends Activity
implements LoaderManager.LoaderCallbacks<Cursor> {
    public static final int MS_LOADER = "ms_crsr".hashCode();
    private MediaCursorAdapter adapter;

    @Override
    protected void onCreate(Bundle savedInstanceState) {
        super.onCreate(savedInstanceState);
        setContentView(R.layout.images);

        adapter = new MediaCursorAdapter(
            getApplicationContext());

        GridView grid = (GridView)findViewById(R.id.grid);
        grid.setAdapter(adapter);

        getLoaderManager().initLoader(MS_LOADER, null, this);
    }

    @Override
    public CursorLoader onCreateLoader(int id, Bundle bundle) {
        return new CursorLoader(this,
            MediaStore.Images.Media.EXTERNAL_CONTENT_URI,
            new String[]{
                MediaStore.Images.Media._ID,
                MediaStore.Images.Media.DISPLAY_NAME
            }, "", null, null);
    }

    @Override
    public void onLoadFinished(
        Loader<Cursor> loader, Cursor media) {
```

```
        adapter.changeCursor(media);
    }

    @Override
    public void onLoaderReset(Loader<Cursor> loader) {
        adapter.changeCursor(null);
    }
}
```

Have a look at the parts in bold in the previous code. We create a MediaCursorAdapter, and pass it to the GridView, we then initialize our CursorLoader. When loading is completed, we pass the loaded Cursor to the Adapter, and we're done.

The remaining piece to implement is MediaCursorAdapter, which is going to start out as a very simple class. The job of our CursorAdapter is simply to map rows of data from the Cursor to each View in the individual GridView cells.

The Android SDK provides the very handy SimpleCursorAdapter class, which does just what we need. So for now we'll just subclass it and instruct it, via constructor parameters, which layout to inflate for each cell and the Cursor fields to map to each View within that layout.

```
public class MediaCursorAdapter extends SimpleCursorAdapter {
    private static String[] FIELDS = new String[]{
        MediaStore.Images.Media._ID,
        MediaStore.Images.Media.DISPLAY_NAME
    };

    private static int[] VIEWS = new int[]{
        R.id.media_id, R.id.display_name
    };

    public MediaCursorAdapter(Context context) {
        super(context, R.layout.example2_cell,
            null, FIELDS, VIEWS, 0);
    }
}
```

The layout files and source code are available on the accompanying website. When you run this Activity, you'll see a two-column grid where each cell contains the ID and display name of an image from the MediaStore. If, like me, you have a lot of images on your device, you can fling the grid and watch the file names spin by.

Scroll to somewhere in the middle of the list and rotate your device, and you'll notice that the Activity restarts and redisplays the grid immediately, without losing its place—this is because the CursorLoader survived the restart, and still holds the Cursor object with the same rows loaded.

This is technically all very interesting, but it isn't much to look at. In the next section we'll combine our two Loaders to implement a scrollable grid displaying the images themselves.

Combining Loaders

In the preceding sections we developed an AsyncTaskLoader that can load a single image thumbnail as a Bitmap, and a CursorLoader that loads a list of all available images on the device. Let's bring them together to create an app that tiles thumbnails of all the images on the device in a scrollable grid, performing all loading in the background.

Thanks to our CursorLoader, we already have access to the IDs of the images we need to load—we're displaying them as text—so we just need to pass those IDs to our ThumbnailLoader for it to asynchronously load the image for us.

Recall that ThumbnailLoader was set up to load one Bitmap and cache it forever (that is, until explicitly removed from LoaderManager). We want to change that so that a single ThumbnailLoader can first be constructed without a mediaId, and later be told to load an image with a particular mediaId.

```
public ThumbnailLoader(Context context) {
    super(context);
}
```

We'll enable ThumbnailLoader to load a new image instead of its current one, by setting a new mediaId. Since the bitmap is cached, just setting a new ID won't suffice—we also need to trigger a reload:

```
public void setMediaId(Integer newMediaId) {
    if ((mediaId != null) && (!mediaId.equals(newMediaId))
        this.mediaId = newMediaId;
        onContentChanged();
    }
}
```

What's this? onContentChanged is a method of the abstract Loader superclass, which will force a background load to occur if our Loader is currently in the "started" state. If we're currently "stopped", a flag will be set and a background load will be triggered next time the Loader is started. Either way, when the background load completes, onLoadFinished will be triggered with the new data.

We need to make a few changes to our onStartLoading method to correctly handle the case where we were "stopped" when onContentChanged was invoked. Let's remind ourselves of what it used to look like:

```
@Override
protected void onStartLoading() {
    if (data != null) {
        deliverResult(data);
    } else {
        forceLoad();
    }
}
```

Now let's update onStartLoading to check if we need to reload our data, and respond appropriately:

```
@Override
protected void onStartLoading() {
    if (data != null)
        deliverResult(data);
    if (takeContentChanged() || data == null)
        forceLoad();
}
```

The updated onStartLoading still delivers its data immediately — if it has any. It then calls takeContentChanged to see if we need to discard our cached Bitmap and load a new one. If takeContentChanged returns true, we invoke forceLoad to trigger a background load and redelivery.

Now we can cause our ThumbnailLoader to load and cache a different image, but a single ThumbnailLoader can only load and cache one image at a time, so we're going to need more than one active instance.

Let's walk through the process of modifying MediaCursorAdapter to initialize a ThumbnailLoader for each cell in the GridView, and to use those Loaders to asynchronously load the images and display them.

First, we can no longer rely on SimpleCursorAdapter, so we'll need to subclass CursorAdapter instead. We'll also need LoaderManager, which we'll pass to the constructor, and we'll grab LayoutInflater from the Context to use later when we create the View objects:

```
public class MediaCursorAdapter extends CursorAdapter {
    private LoaderManager mgr;
    private LayoutInflater inf;
    private int count;
    private List<Integer> ids;

    public MediaCursorAdapter(Context ctx, LoaderManager mgr) {
        super(ctx.getApplicationContext(), null, true);
        this.mgr = mgr;
        inf = (LayoutInflater) ctx.
            getSystemService(Context.LAYOUT_INFLATER_SERVICE);
        ids = new ArrayList<Integer>();
    }

    @Override
    public View newView(
        Context ctx, Cursor crsr, ViewGroup parent) {
        return null; // todo
    }

    @Override
    public void bindView(View view, Context ctx, Cursor crsr) {
        // todo
    }
}
```

We have two methods to implement—newView and bindView. GridView will invoke newView until it has enough View objects to fill all of its visible cells, and from then on it will recycle these same View objects by passing them to bindView to be repopulated with data for a different cell as the grid scrolls. As a view scrolls out of sight, it becomes available for rebinding.

What this means for us is that we have a convenient method in which to initialize our ThumbnailLoaders—newView, and another convenient method in which to retask Loader to load a new thumbnail—bindView.

newView first inflates ImageView for the cell and gives it a unique ID, then passes it to a ThumbnailCallbacks class, which we'll meet in a moment. ThumbnailCallbacks is in turn used to initialize a new Loader, which shares the ID of the View. In this way we are initializing a new Loader for each visible cell in the grid:

```
@Override
public View newView(final Context ctx, Cursor crsr, ViewGroup vg){
    ImageView view = (ImageView)
        inf.inflate(R.layout.example3_cell, vg, false);
    view.setId(
        MediaCursorAdapter.class.hashCode() + count++);
    mgr.initLoader(
        view.getId(), null,
        new ThumbnailCallbacks(ctx, view));
    ids.add(view.getId());
    return view;
}
```

In bindView, we are recycling each existing View to update the image being displayed by that View. So the first thing we do is clear out the old Bitmap.

Next we look up the correct Loader by ID, extract the ID of the next image to load from the Cursor, and load it by passing the ID to a new method of ThumbnailLoader — setMediaId.

```
@Override
public void bindView(View view, final Context ctx, Cursor crsr) {
    ((ImageView)view).setImageBitmap(null);
    ThumbnailLoader loader = (ThumbnailLoader)
        mgr.getLoader(view.getId());
    loader.setMediaId(crsr.getInt(
        crsr.getColumnIndex(MediaStore.Images.Media._ID)));
}
```

We need to add one more method to our Adapter so that we can clean up ThumbnaiLoaders when we no longer need them. We'll call these ourselves when we no longer need these Loaders — for example, when our Activity is finishing:

```
public void destroyLoaders() () {
    for (Integer id : ids)
        mgr.destroyLoader(id);
}
```

That's our completed Adapter. Next, let's look at ThumbnailCallbacks which, as you probably guessed, is just an implementation of LoaderCallbacks:

```
public class ThumbnailCallbacks implements
LoaderManager.LoaderCallbacks<Bitmap> {
```

```
    private Context context;
    private ImageView image;

    public ThumbnailCallbacks(Context context, ImageView image) {
        this.context = context.getApplicationContext();
        this.image = image;
    }

    @Override
    public Loader<Bitmap> onCreateLoader(int i, Bundle bundle) {
        return new ThumbnailLoader(context);
    }

    @Override
    public void onLoadFinished(Loader<Bitmap> loader, Bitmap b) {
        image.setImageBitmap(b);
    }

    @Override
    public void onLoaderReset(Loader<Bitmap> loader) {
    }
}
```

The only interesting things `ThumbnailCallbacks` does is create an instance of `ThumbnailLoader`, and set a loaded `Bitmap` to its `ImageView`.

Our `Activity` is almost unchanged — we need to pass an extra parameter when instantiating `MediaCursorAdapter` and, to avoid leaking the Loaders it creates, we need to invoke the `destroyLoaders` method of `MediaCursorAdapter` in `onPause` or `onStop`, if the `Activity` is finishing:

```
@Override
protected void onStop() {
    super.onStop();

    if (isFinishing()) {
        getSupportLoaderManager().destroyLoader(MS_LOADER);
        adapter.destroyLoaders();
    }
}
```

The full source code is available from the Packt Publishing website. Take a look at the complete source code to appreciate how little there actually is, and run it on a device to get a feel for just how much functionality Loaders give you for a relatively small effort!

Applications of Loaders

Due to its focused nature, applications of Loader are relatively easy to identify. The obvious applications include reading any kind of data from files or databases local to the device, as we've done in the examples in this chapter.

Of course there is no reason that the definition of "loading" should not encompass computing a value or set of values — in the download section of the Packt Publishing website you can find an example that uses Loader to calculate a set of prime numbers.

One strong advantage of Loaders over direct use of AsyncTask is that their lifecycle is very flexible with respect to the Activity and Fragment lifecycles. Without any extra effort we can handle configuration changes such as an orientation change. We can even start loading in one Activity, navigate through the app, and collect the result in a completely separate Activity, if that makes sense for our app.

This decoupling from the Activity lifecycle makes Loader in some ways a better candidate than AsyncTask to perform network transfers such as HTTP downloads; however, they require more code, and still aren't a perfect fit. We'll discuss more appropriate alternatives in *Chapter 5, Queuing Work with IntentService* and *Chapter 6, Long-running Tasks with Service*.

Summary

The Loader framework in Android does a wonderful job of making it easy to load data in the background and deliver it to the main thread when it is ready.

In this chapter we learned about the essential characteristics of all Loaders — background loading, caching of loaded data, and a managed lifecycle.

We took a detailed look at AsyncTaskLoader as a means to perform arbitrary background loading, and CursorLoader for asynchronous loading from local database Cursors.

We saw that Loaders can free us from some of the constraints imposed by the Activity lifecycle, and took advantage of that to continue to work in the background even across Activity restarts.

In the next chapter we'll free ourselves completely from the constraints of the Activity lifecycle and perform long-running background operations with IntentService, even when our app is no longer in the foreground.

5
Queuing Work with IntentService

We're building up quite a toolkit for asynchronous Android programming. We can perform work off the main thread and update the user interface using `HandlerThread` and `Handler`. If we're working in the context of a single Activity and want to show progress as we go, we can use `AsyncTask`. When we need to load data asynchronously and have it survive across Activity restarts, we can turn to Loaders.

What if we want to perform some background work that must complete even if the user exits the application?

Applications usually don't get killed immediately by the system, so we could just start a background thread and hope that the system doesn't kill our application before the work is completed.

However, this would be quite unreliable, and we can be sure that it won't work well on a large percentage of Android devices in the wild. Luckily, there is a solution available in the form of `Service` and its more specialized subclass `IntentService`.

In this chapter, we will cover the following topics:

- Introducing `Service` and `IntentService`
- Building responsive apps with `IntentService`
- Returning results with `PendingIntent`
- Posting results as system notifications
- Applications of `IntentService`
- Reporting progress from `IntentService`

Introducing Service and IntentService

If the basic unit of a visible application is `Activity`, its equivalent unit for non-visible components is `Service`. Just like activities, services must be declared in the `AndroidManifest` file so that the system is aware of them and can manage them for us.

```
<service android:name=".MyService"/>
```

`Service` has lifecycle callback methods similar to those of `Activity`, which are always invoked on the application's main thread.

Also, just like `Activity`, `Service` does not automatically entail a separate background thread or process, and performing intensive or blocking operations in a `Service` callback method can lead to an **Application Not Responding** dialog.

However, there are several ways in which services are different to activities, listed as follows:

- A `Service` does not provide a user interface
- There can be many services active at the same time within an application
- A `Service` can remain active even if the application hosting it is not the current foreground application, which means that there can be many services of many apps all active at the same time
- Because the system is aware of services running within a process, it can avoid killing those processes unless absolutely necessary, allowing the background work to continue

In this chapter, we will focus on the `IntentService` class, a special-purpose subclass of `Service` that makes it very easy to implement a task queue to process work on a single background thread.

Building responsive apps with IntentService

The `IntentService` class is a specialized subclass of `Service` that implements a background work queue using a single `HandlerThread`. When work is submitted to an `IntentService`, it is queued for processing by a `HandlerThread`, and processed in order of submission.

If the user exits the app before the queued work is completely processed, the IntentService will continue working in the background. When the IntentService has no more work in its queue, it will stop itself to avoid consuming unnecessary resources.

The system may still kill a background app with an active IntentService, if it really needs to (to reclaim enough memory to run the current foreground process), but it will kill lower priority processes first, for example, other non-foreground apps that do not have active services.

The `IntentService` class gets its name from the way in which we submit work to it: by invoking `startService` with an Intent.

```
startService(new Intent(context, MyIntentService.class));
```

We can call `startService` as often as we like, which will start the IntentService if it isn't already running, or simply enqueue work to an already running instance if there is one.

If we want to pass some data to a Service, we can do so by supplying a data `Uri` or extra data fields via an Intent.

```
Intent intent = new Intent(context, MyIntentService.class);
intent.setData(uri);
intent.putExtra("param", "some value");
startService(intent);
```

We can create an `IntentService` subclass by extending `android.app.IntentService` and implementing the abstract `onHandleIntent` method. We must invoke the single-argument constructor with a name for its background thread (naming the thread makes debugging and profiling much easier).

```
public class MyIntentService extends IntentService {
  public MyIntentService() {
    super("thread-name");
  }
  protected void onHandleIntent(Intent intent) {
    // executes on the background HandlerThread.
  }
}
```

We'll need to register the IntentService in our `AndroidManifest` file, using a `<service>` element as follows:

```
<service android:name="com.mypackage.MyIntentService"/>
```

If we want our `IntentService` to only be used by the components of our own application, we can specify that it is not `public` with an extra attribute:

```
<service android:name=".MyIntentService"
  android:exported="false"/>
```

Let's get started by implementing `IntentService` to do something we're already familiar with — calculating the *n*th prime:

```
public class PrimesIntentService extends IntentService {
  public static final String PARAM = "prime_to_find";

  public PrimesIntentService() {
    super("primes");
  }

  protected void onHandleIntent(Intent intent) {
    int n = intent.getIntExtra(PARAM, -1);
    BigInteger prime = calculateNthPrime(n);
  }

  private BigInteger calculateNthPrime(int n) {
    BigInteger prime = BigInteger.valueOf(2);
    for (int i=0; i<n; i++)
      prime = prime.nextProbablePrime();
    return prime;
  }
}
```

Notice that we're declaring a `public static` constant name for the parameter, just to make it easy to use the correct name from any client Activity that wants to invoke the Service. We can invoke this `IntentService` to calculate the *n*th prime as follows:

```
private void triggerIntentService(int primeToFind) {
  Intent intent = new Intent(this, PrimesIntentService.class);
  intent.putExtra(PrimesIntentService.PARAM, primeToFind);
  startService(intent);
}
```

So far so good, but you've probably noticed that we haven't done anything with the result we calculated. In the next section, we'll look at some of the ways in which we can send results from services to activities or fragments.

Handling results

Any `Service` — including subclasses of `IntentService` — can be used to start background work from which the originating `Fragment` or `Activity` doesn't expect a response.

However, it is very common to need to return a result or display the result of the background work to the user. We have several options for doing this:

- Send a `PendingIntent` to the Service from the originating Activity, allowing the Service to callback to the Activity via its `onActivityResult` method
- Post a system notification allowing the user to be informed that the background work was completed, even if the application is no longer in the foreground
- Send a `Message` to a `Handler` in the originating Activity using Messenger
- Broadcast the result as an `Intent`, allowing any `Fragment` or `Activity` — including the originator — to receive the result of background processing

We'll learn about `Messenger` and `BroadcastReceiver` in *Chapter 6, Long-running Tasks with Service*. In this chapter, we'll return results with `PendingIntent`, and alert the user with system notifications.

Returning results with PendingIntent

When we invoke an `IntentService`, it does not automatically have any way to respond to the calling `Activity`; so if the Activity wants to receive a result, it must provide some means for the `IntentService` to reply.

Arguably, the easiest way to do that is with `PendingIntent`, which will be familiar to any Android developer who has worked with multiple activities using the `startActivityForResult` and `onActivityResult` methods, as the pattern is essentially the same.

First, we'll add a few static members to `PrimesIntentService` to ensure that we use consistent values between it and the calling `Activity`:

```
public static final String PENDING_RESULT = "pending_result";
public static final String RESULT = "result";
public static final int RESULT_CODE = "nth_prime".hashCode();
```

We'll also need to define a static member in our `Activity` for the `REQUEST_CODE` constant, which we can use to correctly identify the results returned to our `onActivityResult` method:

```
private static final int REQUEST_CODE = 0;
```

Now, when we want to invoke `PrimesIntentService` from our `Activity`, we'll create a `PendingIntent` for the current `Activity`, which will act as a callback to invoke its `onActivityResult` method.

We can create a `PendingIntent` with the `createPendingResult` method of `Activity`, which accepts three parameters: an `int` result code, an `Intent` to use as the default result, and an `int` that encodes some configuration for how the `PendingIntent` can be used (for example, whether it may be used more than once).

```
PendingIntent pending = createPendingResult(
  REQUEST_CODE, new Intent(),(), 0);
```

We pass the `PendingIntent` to the `IntentService` by adding it as an *extra* in the Intent we launch the `IntentService` with:

```
private void triggerIntentService(int primeToFind) {
  PendingIntent pending = createPendingResult(
    REQUEST_CODE, new Intent(), 0);
  Intent intent = new Intent(this, PrimesIntentService.class);
  intent.putExtra(PrimesIntentService.PARAM, primeToFind);
  intent.putExtra(
  PrimesIntentService.PENDING_RESULT, pending);
  startService(intent);
}
```

To handle the result that will be returned when this `PendingIntent` is invoked, we need to implement `onActivityResult` in the Activity, and check for the result code:

```
protected void onActivityResult(int req, int res, Intent data) {
  if (req == REQUEST_CODE &&
      res == PrimesIntentService.RESULT_CODE) {
    BigInteger result = (BigInteger)
      data.getSerializableExtra(PrimesIntentService.RESULT);
    // … update UI with the result
  }
  super.onActivityResult(requestCode, resultCode, data);
}
```

The IntentService can now reply to the calling Activity by invoking one of the PendingIntent's send methods with the appropriate request code. Our updated onHandleIntent method looks as follows:

```
protected void onHandleIntent(Intent intent) {
    int n = intent.getIntExtra(PARAM, -1);
    BigInteger prime = calculateNthPrime(n);
    try {
      Intent result = new Intent();
      result.putExtra(RESULT, prime);
      PendingIntent reply =
        intent.getParcelableExtra(PENDING_RESULT);
        reply.send(this, RESULT_CODE, result);
    } catch (PendingIntent.CanceledException exc) {
      Log.i(TAG, "reply cancelled", exc);
    }
}
```

The additional code creates a new Intent object and populates it with our calculated prime, then sends it back to the calling Activity using PendingIntent. We also have to handle the CanceledException, in case the calling Activity decided that it wasn't interested in the result any more and cancelled the PendingIntent. That's all there is to it—our Activity will now be invoked via its onActivityResult method when the IntentService completes its work.

As a bonus, we will still receive the result if the Activity has restarted, for example, due to configuration changes such as a device rotation.

What if the user left the Activity (or even left the application) while the background work was in progress? In the next section, we'll use notifications to provide feedback without interrupting the user's new context.

Posting results as system notifications

System notifications appear initially as an icon in the notification area, normally at the very top of the device screen. Once notified, the user can open the notification drawer to see more details.

Notifications are an ideal way to inform the user of results or status updates from services, particularly when the operation may take a long time to complete and the user is likely to be doing something else in the meantime.

Let's post the result of our prime number calculation as a notification, with a message containing the result that the user can read when they open the notification drawer. We'll use the support library to ensure broad API level compatibility, and add one method to `PrimesIntentService` as follows:

```
private void notifyUser(int primeToFind, String result) {
  String msg = String.format(
    "The %sth prime is %s", primeToFind, result);
  NotificationCompat.Builder builder =
    new NotificationCompat.Builder(this)
      .setSmallIcon(R.drawable.prime_notification_icon)
      .setContentTitle(getString(R.string.primes_app))
      .setContentText(msg);
  NotificationManager nm = (NotificationManager)
    getSystemService(Context.NOTIFICATION_SERVICE);
  nm.notify(primeToFind, builder.build());
}
```

Here we're using `NotificationCompat.Builder` to build a notification that includes an icon, a title (just the name of our application), and a message containing the result of the calculation.

Each notification has an identifier, which we can use to control whether a new notification is posted or an existing one is reused. The identifier is an `int`, and is the first parameter to the `notify` method. Since our `primeToFind` value is an `int`, and we would like to be able to post multiple notifications, it makes sense to use `primeToFind` as the ID for our notifications so that each different request can produce its own separate notification.

To post a notification containing the result of our calculation, we just need to update `onHandleIntent` to invoke the `notifyUser` method:

```
protected void onHandleIntent(Intent intent) {
  int n = intent.getIntExtra(PARAM, -1);
      BigInteger prime = calculateNthPrime(n);
  notifyUser(n, prime.toString());
}
```

Now that we've learned the basics of using `IntentService`, let's consider some real-world applications.

Applications of IntentService

Ideal applications for IntentService include just about any long-running task where the work is not especially tied to the behavior of a Fragment or Activity, and particularly when the task must complete its processing regardless of whether the user exits the application.

However, IntentService is only suitable for situations where a single worker thread is sufficient to handle the workload, since it's work is processed by a single HandlerThread, and we cannot start more than one instance of the same IntentService subclass.

A usecase that IntentService is ideally suited for is uploading data to remote servers. An IntentService is a perfect fit because:

- The upload usually must complete, even if the user leaves the application
- A single upload at a time usually makes best use of the available connection, since bandwidth is often asymmetric (there is much smaller bandwidth for upload than download)
- A single upload at a time gives us a better chance of completing each individual upload before losing our data connection

Let's see how we might implement a very simple IntentService that uploads images from the MediaStore to a simple web service via HTTP POST.

HTTP uploads with IntentService

For this example, we'll create a new Activity, UploadPhotoActivity, to allow the user to pick an image to upload. We'll start from the code for MediaStoreActivity that we created in *Chapter 4, Asynchronous I/O with Loader*.

Our new UploadPhotoActivity only needs a small modification to add an OnItemClickListener interface to the GridView of images, so that tapping an image triggers its upload. We can add the listener as an anonymous inner class in onCreate as follows:

```
grid.setOnItemClickListener(
  new AdapterView.OnItemClickListener()) {
  public void onItemClick(
    AdapterView<?> parent, View view, int position, long id) {
    Cursor cursor = (Cursor)adapter.getItem(position);
    int mediaId = cursor.getInt(
```

```
          cursor.getColumnIndex(MediaStore.Images.Media._ID));
      Uri uri = Uri.withAppendedPath(
        MediaStore.Images.Media.EXTERNAL_CONTENT_URI,
        Integer.toString(mediaId));
      Intent intent = new Intent(
        UploadPhotoActivity.this, UploadIntentService.class);
      intent.setData(uri);
      startService(intent);
    }
  });
```

This looks like quite a dense chunk of code, but all it really does is use the position of the tapped thumbnail to move the Cursor to the correct row in its result set, extract the ID of the image that was tapped, create a Uri for its original file, and then start UploadIntentService with an Intent containing that Uri.

We'll extract the details of the upload into a separate class, so UploadIntentService itself is just a fairly sparse IntentService implementation. In onCreate, we'll set up an instance of our ImageUploader class, which will be used to process all uploads added to the queue during this lifetime of UploadIntentService.

```
public void onCreate() {
  super.onCreate();
  uploader = new ImageUploader(getContentResolver());
}
```

The onHandleIntent method just does the coordination duties of notifying the user that an upload is in progress, using an ImageUploader instance to perform the upload, then notifying the user of success or failure.

```
protected void onHandleIntent(Intent intent) {
  Uri data = intent.getData();
  int id = Integer.parseInt(data.getLastPathSegment());
  sendNotification(id, String.format("Uploading %s.jpg",id));
  if (uploader.upload(data, null)) {
    notifyUser(id, String.format("Completed %s.jpg",id));
  } else {
    notifyUser(id, String.format("Failed %s.jpg",id));
  }
}
```

The implementation of ImageUploader itself is not all that interesting — we just use Java's HTTPURLConnection class to post the image data to the server. The complete source code is available on the Packt Publishing website, so we'll just list two critical methods — upload and pump — and leave out the housekeeping.

```
public boolean upload(Uri data, ProgressCallack callback) {
  HttpURLConnection conn = null;
  try {
    int len = getContentLength(data);
    URL destination = new URL(UPLOAD_URL);
    conn = (HttpURLConnection) destination.openConnection();
    conn.setDoInput(true);
    conn.setDoOutput(true);
    conn.setFixedLengthStreamingMode(len);
    conn.setRequestMethod("POST");
    conn.setRequestProperty("Content-Type", "image/jpg");
    conn.setRequestProperty("Content-Length", len + "");
    conn.setRequestProperty(
      "Filename", data.getLastPathSegment() + ".jpg");
    InputStream in = null;
    OutputStream out = null;
    try {
      pump(
        in = content.openInputStream(data),
        out = conn.getOutputStream(),
        callback, len);
    } finally {
        close(in);
        close(out);
    }

    return ((conn.getResponseCode() >= 200) &&
        (conn.getResponseCode() < 400));
    }
    catch (IOException exc) {
      Log.e(TAG, "upload failed", exc);
      return false;
    } finally {
      conn.disconnect();
    }
  }
}
```

The pump method just copies 1 KB chunks of data from the InputStream to the OutputStream, pumping the data to the server, as follows:

```
private void pump(
  InputStream in, OutputStream out,
    ProgressCallback callback, int len)
    throws IOException {
```

```
int length,i=0,size=1024;
byte[] buffer = new byte[size]; // 1kb buffer
while ((length = in.read(buffer)) > -1) {
    out.write(buffer, 0, length);
    out.flush();
    if (callback != null)
      callback.onProgress(len, ++i*size);
}
}
```

Each time a 1 KB chunk of data is pushed to the `OutputStream`, we invoke the `ProgressCallback` method, which we'll use in the next section to report the progress to the user.

Reporting progress

For long-running processes, it can be very useful to report progress so that the user can take comfort in knowing that something is actually happening.

To report progress from an `IntentService`, we can use the same mechanisms that we use to send results — for example, sending `PendingIntents` containing progress information, or posting system notifications with progress updates.

We can also use some techniques that we'll cover in the next chapter: sending messages via `Messenger`, or broadcasting intents to registered receivers.

 Whichever approach we take to report progress, we should be careful not to report progress too frequently, otherwise we'll waste resources updating the progress bar at the expense of completing the work itself!

Let's look at an example that displays a progress bar on notifications in the drawer — a use case that the Android development team anticipated and therefore made easy for us with the `setProgress` method of `NotificationCompat.Builder`:

```
Builder setProgress(int max, int progress, boolean indeterminate);
```

Here, `max` sets the target value at which our work will be completed, `progress` is where we have got to so far, and `indeterminate` controls which type of progress bar is shown. When `indeterminate` is `true`, the notification shows a progress bar that indicates something is in progress without specifying how far through the operation we are, while `false` shows the kind of progress bar that we need — one that shows how much work we have done, and how much is left to do.

We'll need to calculate progress and dispatch notifications at appropriate intervals, which we've facilitated through our `ProgressCallback` class. Now we need to implement the `ProgressCallback`, and hook it up in `UploadIntentService`:

```
private class ProgressNotificationCallback
implements ImageUploader.ProgressCallback {

  private NotificationCompat.Builder builder;
  private NotificationManager nm;
  private int id, prev;

  public ProgressNotificationCallback(
    Context ctx, int id, String msg) {
    this.id = id;
    prev = 0;
    builder = new NotificationCompat.Builder(ctx)
      .setSmallIcon(android.R.drawable.stat_sys_upload_done)
      .setContentTitle(getString(R.string.upload_service))
      .setContentText(msg)
      .setProgress(100,0,false);
    nm = (NotificationManager)
        getSystemService(Context.NOTIFICATION_SERVICE);
    nm.notify(id, builder.build());
  }

  public void onProgress(int max, int progress) {
    int percent = (int)((100f*progress)/max);
    if (percent > (prev + 5))))))))) {
      builder.setProgress(100, percent, false);
      nm.notify(id, builder.build());
      prev = percent;
    }
  }

  public void onComplete(String msg) {
    builder.setProgress(0, 0, false);
    builder.setContentText(msg);
    nm.notify(id, builder.build());
  }
}
```

The constructor of `ProgressNotificationCallback` consists of the familiar code to post a notification with a progress bar.

The onProgress method throttles the rate at which notifications are dispatched, so that we only post an update as each additional 5 percent of the total data is uploaded—in order not to swamp the system with notification updates.

The onComplete method posts a notification that sets both the integer progress parameters to zero, which removes the progress bar.

The final change is to make onHandleIntent use ProgressNotificationCallback as follows:

```
protected void onHandleIntent(Intent intent) {
  Uri data = intent.getData();
  // unique id per upload, so each has its own notification
  int id = Integer.parseInt(data.getLastPathSegment());
  String msg = String.format("Uploading %s.jpg",id);
  ProgressNotificationCallback progress =
    new ProgressNotificationCallback(this, id, msg);
  if (uploader.upload(data, progress)) {
    progress.onComplete(
      String.format("Uploaded %s.jpg", id));
  } else {
    progress.onComplete(
      String.format("Upload failed %s.jpg", id));
  }
}
```

If you download the sample code from the Packt Publishing website, you can try to upload an image from your phone to a simple web service running on Google App Engine at http://devnullupload.appspot.com/upload (don't worry, the web service just consumes the bytes and throws them away, so your images are safe!)

Tap an image to start uploading and you'll see a notification appear. Slide open the notification drawer and watch the progress bar ticking up in 5 percent increments as your image uploads.

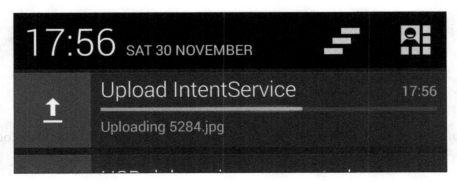

Summary

In this chapter, we explored the incredibly useful `IntentService` — an ideal construct for performing long-running background tasks off the main thread, surviving well beyond the lifecycle of the initiating Activity, and even continuing to do useful work when the application is no longer in the foreground.

We learned how to send work to an `IntentService` with parameterized Intents, how to process that work in the background by implementing `onHandleIntent`, and how to send results back to the originating Activity using a `PendingIntent`.

For cases where the application is no longer in the foreground or an operation is particularly long-running, we saw how to post notifications to the notification drawer, complete with progress updates.

In the next chapter, we'll use IntentService's superclass — `Service` — to perform work on multiple background threads even when the host application is not in the foreground.

6
Long-running Tasks with Service

In *Chapter 5, Queuing Work with IntentService*, we learned about a specialized subclass of `Service` that handles its workload on a single background thread.

In this chapter, we'll extend our toolkit by directly extending `Service` to take control of the level of concurrency applied to our long-running background tasks—how many threads are used to perform the work—and use various methods to send work to Services and receive results from them.

In this chapter, we will cover the following topics:

- Building responsive apps with `Service`
- Controlling concurrency with Executors
- Returning results with `Messenger`
- Direct communication with local Services
- Broadcasting results with Intents
- Detecting unhandled broadcasts
- Applications of Services

Building responsive apps with Service

Throughout this book, we have learned about the concurrency constructs provided by the Android platform for doing work off the main thread. So, it might seem surprising that, by itself, `Service` does not provide any background threads and will run all of its callbacks directly on the main thread, just like `Activity`.

If we perform long-running work or block the main thread in a `Service` callback method, our application may be shut down, and a system-triggered **Application Not Responding** dialog is presented to the user.

While it is possible to configure the service to launch in a separate process, that process will still run the `Service` callbacks on its own main thread and will be subject to the same constraints. The only difference is that our foreground process will not be shut down along with the misbehaving `Service` process.

The solution, of course, is to pass the work off from the main thread to background "worker" threads.

`IntentService`, which we learned about in the previous chapter, does exactly this by passing work from the main thread to a single background `HandlerThread`—an elegant design and a nice example of using the platform concurrency constructs as building blocks to compose concurrent applications.

Elegant though it may be, we can only invoke `IntentService` via an Intent and it will queue all work and process it on a single thread. These design choices make `IntentService` easy to use, but can also be limiting.

When we need more control over the level of concurrency, or alternative methods to trigger long-running background work, we can create our own `Service` implementations and take complete control.

Controlling concurrency with Executors

The `Executor` interface was introduced to the core Java libraries in Java 5 as a means of submitting tasks to be executed without specifying exactly how or when the execution will be carried out. We learned about `Executor` briefly in *Chapter 2, Staying Responsive with AsyncTask*, and used it to control the level of concurrency of our AsyncTasks.

In this section, we'll use Executor to create our own alternative to IntentService, initiating background work in a Service by sending it an Intent, but retaining control over the level of concurrency. Like IntentService, our class will be abstract and should be subclassed to define the actual behavior. So, we'll define an abstract onHandleIntent method for the subclasses to implement:

```java
public abstract class ConcurrentIntentService extends Service {
    protected abstract void onHandleIntent(Intent intent);
}
```

We must override the onBind method of Service, but for now, we will return null, as we won't be using it just yet:

```java
@Override
public IBinder onBind(Intent intent) {
    return null;
}
```

We'll allow subclasses to define the level of concurrency by passing an Executor to the constructor.

```java
public abstract class ConcurrentIntentService extends Service {
    private final Executor executor;
    public ConcurrentIntentService(Executor executor) {
        this.executor = executor;
    }

    protected abstract void onHandleIntent(Intent intent);
}
```

The onHandleIntent method will be invoked in the background using the Executor and triggered from the onStartCommand method of our Service, which is invoked by the platform when we call startService from our Activity.

```java
@Override
public int onStartCommand(
    final Intent intent, int flags, int startId) {
    executor.execute(new Runnable(){
        @Override
        public void run() {
            onHandleIntent(intent);
        }
    });
    return START_REDELIVER_INTENT;
}
```

Note that we're returning START_REDELIVER_INTENT from onStartCommand, which tells the system that if it must kill our Service, for example, to free up memory for a foreground application, it should be scheduled to restart when the system is under less pressure and the last Intent object sent to the Service should be redelivered to it.

Another sensible return value here might be START_NOT_STICKY, which would mean the Service is not automatically restarted and Intents are not redelivered. The third common flag — START_STICKY — is not appropriate to replicate the behavior of IntentService, because we don't want to restart this Service without an Intent to process.

That's all we really need to do to implement our concurrent Intent-driven Service, but we should be responsible and stop the Service when it has no more work to do. We'll need to keep track of how many tasks are running at any given time and when the last one completes, invoke stopSelf.

To track tasks, we'll add a simple int counter property to the class and increment it immediately when we receive a new Intent. We'll use a Handler created on the main thread to send messages from the background threads to the main thread when a task completes. This Handler will decrement the counter again and stop the Service when the count reaches zero. By doing all of the tracking in a Handler on the main thread, we avoid any synchronization issues.

```
private final CompletionHandler handler =
  new CompletionHandler();
private int counter;

@Override
public void onStart(final Intent intent, int startId) {
  counter++;
  executor.execute(new Runnable(){
      @Override
      public void run() {
        try {
          onHandleIntent(intent);
        } finally {
          handler.sendMessage(Message.obtain(handler));
        }
      }
  });
}

private class CompletionHandler extends Handler {
  @Override
```

```
public void handleMessage(Message msg) {
  if (--counter == 0) {
    Log.i(TAG, "0 tasks, stopping");
    stopSelf();
  } else {
    Log.i(TAG, counter + " active tasks");
  }
}
}
```

We can now use our multithreaded Intent-driven Service exactly as we used IntentService. We subclass it, this time passing an Executor configured to run tasks in as many threads as we need, then invoke it from our Activities and Fragments using Intents.

Using our new ConcurrentIntentService, we can make some very small modifications to PrimesIntentService from *Chapter 5, Queuing Work with IntentService*, to have it perform its calculations using a pool of threads:

```
public class PrimesIntentService extends ConcurrentIntentService {
  public static final int MAX_CONCURRENCY = 5;
  public static final String PARAM = "prime_to_find";

  public PrimesIntentService() {
    super(Executors.newFixedThreadPool(
      MAX_CONCURRENCY,
      new ThreadFactory(){
        @Override
        public Thread newThread(Runnable r) {
          Thread t = new Thread(r);
          t.setPriority(Thread.MIN_PRIORITY);
          t.setName("primes-intent-thread");
          return t;
        }
      }));
    }

  // … the rest of the class is unchanged from chapter 5
}
```

The changes we made are very small. We now extend ConcurrentIntentService instead of IntentService, and we used a factory method from the Executor JDK class to create a fixed-size pool of threads to run our tasks with.

Note that we supplied a `ThreadFactory` and used it to set the priority of the threads in our pool to `Thread.MIN_PRIORITY`. It is crucial to do this so that our background worker threads don't starve the main thread of CPU time.

> An early warning sign that our worker threads are contending too heavily with the main thread is a log message like this:
>
> ```
> I/Choreographer: Skipped 53 frames!  The
> application may be doing too much work on its
> main thread.
> ```
>
> Despite the message accusing the application of doing too much on the main thread, it may also be produced if the application is doing too much work in worker threads which have too high a priority, and as a result compete too strongly with the main thread for CPU time.

We can invoke our new concurrent version of `PrimesIntentService` just as we invoked the single-threaded version in the previous chapter by calling `startService` with an `Intent`:

```
Intent intent = new Intent(this, PrimesIntentService.class);
intent.putExtra(PrimesIntentService.PARAM, primeToFind);
startService(intent);
```

Returning results with Messenger

When we're doing work in a Service on behalf of an Activity, it is very common to want to send results back to that Activity, even across configuration changes that cause restarts. Wouldn't it be ideal if we could give our `PrimesIntentService` a Handler that it can use to send messages for processing in the context of the Activity? The great news is we can!

The Android framework provides the `Messenger` class, which wraps up `Handler` and makes it possible to send messages from anywhere—including from remote Services in other processes. `Messenger` implements `Parcelable`, which means we can pass Messengers around in Intents, which we can't do with `Handler` directly.

Let's update our `Service` and `Activity` to communicate results using `Messenger`!

In `PrimesIntentService`, we'll add a `static final int` member to ensure that we use consistent values for the `message.what` property and a `static final String` member to use as the Intent's extra key when passing our `Messenger`.

```
public static final int RESULT = "nth_prime".hashCode();
public static final String MSNGR = "messenger";
```

The first thing we'll need in the Activity is a `Handler` subclass that can process the result messages. We'll define it as a static inner class of our Activity, so that we don't accidently leak implicit references to the enclosing class.

```
private static class PrimesHandler extends Handler {
  private TextView view;
  public void handleMessage(Message message) {
    if (message.what == PrimesIntentService.RESULT) {
      if (view != null) { // if we're attached
        view.setText(message.obj.toString());
      }
    }
  }
  public void attach(TextView view) {
    this.view = view;
  }
  public void detach() {
    this.view = null;
  }
}
```

Our `Handler` member is also declared `static` so the same instance is available across Activity restarts, and uses the attach/detach pattern we first saw in *Chapter 3, Distributing Work with Handler and HandlerThread*, to ensure we don't leak references to the `View` hierarchy. We'll also create a `static` `Messenger` in our `Activity` that wraps our `Handler`.

```
private static PrimesHandler handler = new PrimesHandler();
private static Messenger messenger = new Messenger(handler);

@Override
protected void onStart() {
    super.onStart();
    handler.attach((TextView) findViewById(R.id.results));
}

@Override
protected void onStop() {
    super.onStop();
    handler.detach();
}
```

We can pass our `Messenger` to `PrimesIntentService` via an `Intent` that initiates a calculation.

```
Intent intent = new Intent(this, PrimesIntentService.class);
intent.putExtra(PrimesIntentService.PARAM, primeToFind);
intent.putExtra(PrimesIntentService.MSNGR, messenger);
startService(intent);
```

The `onHandleIntent` method of `PrimesIntentService` will need to extract the Messenger from the Intent and use it to send results back to the Activity. Sending messages with `Messenger` is very similar to sending them with `Handler` — we obtain a `Message` with the appropriate parameters and then send it with `Messenger.send`:

```
@Override
protected void onHandleIntent(Intent intent) {
  int primeToFind = intent.getIntExtra(PARAM, -1);
  Messenger messenger = intent.getParcelableExtra(MSNGR);
  try {
    if (primeToFind < 2) {
      messenger.send(Message.obtain(null, INVALID));
    } else {
      messenger.send(Message.obtain(
        null, RESULT, primeToFind, 0,
        calculateNthPrime(primeToFind)));
    }
  } catch (RemoteException anExc) {
    Log.e(TAG, "Unable to send message", anExc);
  }
}
```

In this section, we've started background work in a `Service` using Intents, and given the `Service` a communication channel it can use to send results back to the `Activity`, even if the `Activity` is restarted. In the next section, we'll look at some alternative ways to initiate and communicate with Services.

Communicating with Services

In all of our dealings with Services so far, we have initiated work by invoking `startService` with an `Intent`, but that isn't our only option. If our `Service` is designed to only be used locally from within our own application process, we can take significant shortcuts and work with `Service` just as we do with any other Java object.

Direct communication with local Services

To create a `Service` that we can interact with directly, we must implement the `onBind` method that we previously ignored and from which we returned `null`. This time, we'll return an implementation of `IBinder` that provides direct access to the Service it binds. We'll always return the same `IBinder` as shown in the following code:

```java
public class LocalPrimesService extends Service {
  public class Access extends Binder {
    public LocalPrimesService getService() {
      return LocalPrimesService.this;
    }
  };
  private final Access binder = new Access();

  @Override
  public IBinder onBind(Intent intent) {
    return binder;
  }
}
```

An `Activity` or `Fragment` that wants to directly interact with this Service needs to bind to it using the `bindService` method and supply a `ServiceConnection` to handle the connect/disconnect callbacks.

Services such as `LocalPrimesService` that are started by a client binding to them are known as "bound" Services, and stop themselves automatically when all clients have unbound. By contrast, our `ConcurrentIntentService` is a "started" Service. A Service can be both "bound" and "started," but a Service that is explicitly started must also be explicitly stopped.

The `ServiceConnection` implementation simply casts the `IBinder` it receives to the concrete class defined by the `Service`, obtains a reference to the `Service`, and records it in a field of the `Activity`.

```java
public class LocalPrimesActivity extends Activity {
  private LocalPrimesService service;
  private ServiceConnection connection;

  private class Connection implements ServiceConnection {
    @Override
    public void onServiceConnected(
      ComponentName name, IBinder binder) {
```

```
            LocalPrimesService.Access access =
              ((LocalPrimesService.Access)binder);
            service = access.getService();
        }

        @Override
        public void onServiceDisconnected(ComponentName name) {
            service = null;
        }
    }
}
```

We can make the Activity bind and unbind during its onResume and onPause lifecycle methods:

```
@Override
protected void onResume() {
    super.onResume();
    bindService(
        new Intent(this, LocalPrimesService.class),
        connection = new Connection(),
        Context.BIND_AUTO_CREATE);
}

@Override
protected void onPause() {
    super.onPause();
    unbindService(connection);
}
```

This is great—once the binding is made, we have a direct reference to the Service instance and can call its methods! However, we didn't implement any methods in our Service yet, so it's currently useless. Let's add a method to LocalPrimesService that calculates the *nth* prime number in the background using AsyncTask:

```
public void calculateNthPrime(final int n) {
    new AsyncTask<Void,Void,BigInteger>() {
        @Override
        protected BigInteger doInBackground(Void... params) {
            BigInteger prime = new BigInteger("2");
            for (int i=0; i<n; i++) {
                prime = prime.nextProbablePrime();
            }
            return prime;
        }
```

```
      @Override
      protected void onPostExecute(BigInteger result) {
        // todo—communicate result to user
      }
    }.execute();
}
```

Since we now have a direct object reference to `LocalPrimesService` in our `LocalPrimesActivity`, we can go ahead and invoke its `calculateNthPrime` method directly—taking care to check that the `Service` is actually bound first, of course.

```
if (service != null) {
  service.calculateNthPrime(500);
}
```

This is a very convenient and efficient way of submitting work to a `Service`—there's no need to package up a request in an `Intent`, so there's no excess object creation or communication overhead.

Since `calculateNthPrime` is asynchronous, we can't return a result directly from the method invocation, and `LocalPrimesService` itself has no user interface, so how can we present results to our user?

One possibility is to pass a callback to `LocalPrimesService` so that we can invoke methods of our Activity when the background work completes. In `LocalPrimesService`, we define a callback interface for the Activity to implement:

```
public interface Callback {
  public void onResult(BigInteger result);
}
```

There is a serious risk that by passing an `Activity` into the `Service`, we'll expose ourselves to memory leaks. The lifecycles of `Service` and `Activity` do not coincide, so strong references to an `Activity` from a `Service` can prevent it from being garbage collected in a timely fashion.

The simplest way to prevent such memory leaks is to make sure that `LocalPrimesService` only keeps a weak reference to the calling `Activity` so that when its lifecycle is complete, the Activity can be garbage collected, even if there is an ongoing calculation in the Service. The modified `calculateNthPrime` method of `LocalPrimesService` is shown in the following code:

```
public void calculateNthPrime(final int n, Callback activity) {
  final WeakReference<Callback> maybeCallback =
```

```
        new WeakReference<Callback>(activity);
    new AsyncTask<Void,Void,BigInteger>(){
      @Override
      protected BigInteger doInBackground(Void... params) {
        BigInteger prime = new BigInteger("2");
        for (int i=0; i<n; i++) {
          prime = prime.nextProbablePrime();
        }
        return prime;
      }

      @Override
      protected void onPostExecute(BigInteger result) {
        Callback callback = maybeCallback.get();
        if (callback != null)
          callback.onResult(result);
      }
    }.execute();
}
```

We invoke the callback on the main thread using onPostExecute, so that PrimesActivity can interact with the user interface directly in the callback method. We can implement the callback as a method of PrimesActivity:

```
public class PrimesActivity extends Activity
implements LocalPrimesService.Callback {

  @Override
  public void onResult(BigInteger result) {
        // … display the result
  }

  // other methods elided for brevity…
}
```

Now we can directly invoke methods on LocalPrimesService and return results via a callback method of PrimesActivity by passing the Activity itself as the callback:

```
if (service != null) {
  service.calculateNthPrime(500, this);
}
```

This direct communication between `PrimesActivity` and `LocalPrimesService` is very efficient and easy to work with. However, there is a downside: if the Activity restarts because of a configuration change, such as a device rotation, the `WeakReference` to the callback will be garbage collected and `LocalPrimesService` cannot send the result.

If we want to retain the ability to respond, even across `Activity` restarts, we'll need a communication channel that can be reattached to our restarted `Activity` instance.

Earlier, we saw that we can use a `Messenger` to communicate results from a `Service` to the originating Activity, even across Activity restarts. When dealing with bound local Services, we can skip `Messenger` and directly use a `Handler`:

```
public void calculateNthPrime(final int n, Handler handler);
```

This is a very efficient option when we only care about collecting the result in the originating Activity. If we want to allow other parts of the application to also receive the results, we need a different mechanism — broadcasts.

Broadcasting results with Intents

Broadcasting an Intent is a way of sending results to anyone who registers to receive them. This can even include other applications in separate processes if we choose, but if the Activity and Service are a part of the same process, broadcasting is best done using a local broadcast, as this is more efficient and secure.

We can update `LocalPrimesService` to broadcast its results with just a few extra lines of code. First, let's define two constants to make it easy to register a receiver for the broadcast and extract the result from the broadcast `Intent` object:

```
public static final String PRIMES_BROADCAST =
  "com.packt.PRIMES_BROADCAST";
public static final String RESULT = "nth_prime";
```

Now we can implement the method that does most of the work using the `LocalBroadcastManager` to send an `Intent` object containing the calculated result. We're using the support library class `LocalBroadcastManager` here for efficiency and security — broadcasts sent locally don't incur the overhead of interprocess communication and cannot be leaked outside of our application.

```
private void broadcastResult(String result) {
  Intent intent = new Intent(PRIMES_BROADCAST);
  intent.putExtra(RESULT, result);
```

```
      LocalBroadcastManager.getInstance(this).
        sendBroadcast(intent);
  }
```

The `sendBroadcast` method is asynchronous and will return immediately without waiting for the message to be broadcast and handled by receivers. Finally, we invoke our new `broadcastResult` method from `calculateNthPrime`:

```
  @Override
  public void calculateNthPrime(final int n) {
    new AsyncTask<Void,Void,Void>(){
      @Override
      protected Void doInBackground(Void... params) {
        BigInteger prime = new BigInteger("2");
        for (int i=0; i<n; i++)
          prime = prime.nextProbablePrime();
        broadcastResult(prime.toString());
        return null;
        }
    }.execute();
  }
```

Great! We're broadcasting the result of our background calculation. Now we need to register a receiver in `PrimesActivity` to handle the result. Here's how we might define our `BroadcastReceiver` subclass:

```
  private static class NthPrimeReceiver extends BroadcastReceiver {
    private TextView view;

    @Override
    public void onReceive(Context context, Intent intent) {
      if (view != null) {
        String result = intent.getStringExtra(
          LocalPrimesService.RESULT);
        view.setText(result);
      } else {
        Log.i(TAG, " ignoring - we're detached");
      }
    }

    public void attach(TextView view) {
      this.view = view;
    }
```

```
    public void detach() {
      this.view = null;
    }
}
```

This `BroadcastReceiver` implementation is quite simple—all it does is extract and display the result from the `Intent` it receives—basically fulfilling the role of the `Handler` we used in the previous section.

We only want this `BroadcastReceiver` to listen for results while our `Activity` is at the top of the stack and visible in the application, so we'll register and unregister it in the `onStart` and `onStop` lifecycle methods. As with the `Handler` that we used previously, we'll also apply the attach/detach pattern to make sure we don't leak `View` objects:

```
private NthPrimeReceiver receiver = new NthPrimeReceiver();

@Override
protected void onStart() {
  super.onStart();
  bindService(
    new Intent(this, LocalPrimesService.class),
    connection = new Connection(),
    Context.BIND_AUTO_CREATE);
  receiver.attach((TextView)
    findViewById(R.id.result));
  IntentFilter filter = new IntentFilter(
    LocalPrimesService.PRIMES_BROADCAST);
  LocalBroadcastManager.getInstance(this).
    registerReceiver(receiver, filter);
}

@Override
protected void onStop() {
  super.onStop();
  unbindService(connection);
  LocalBroadcastManager.getInstance(this).
    unregisterReceiver(receiver);
  receiver.detach();
}
```

Of course, if the user moves to another part of the application that doesn't register a `BroadcastReceiver`, or if we exit the application altogether, they won't see the result of the calculation.

If our `Service` could detect unhandled broadcasts, we could modify it to alert the user with a system notification instead. We'll see how to do that in the next section.

Detecting unhandled broadcasts

In the previous chapter, we used system notifications to post results to the notification drawer—a nice solution when the user has navigated away from our app before the background work has completed. However, we don't want to annoy the user by posting notifications when our app is still in the foreground and can display the results directly.

Ideally, we'll display the results in the app if it is still in the foreground and send a notification otherwise. If we're broadcasting results, the Service will need to know if anyone handled the broadcast and if not, send a notification.

One way to do this is to use the sendBroadcastSync synchronous broadcast method and take advantage of the fact that the Intent object we're broadcasting is mutable (any receiver can modify it). To begin with, we'll add one more constant to LocalIntentService:

```
public static final String HANDLED = "intent_handled";
```

Next, modify broadcastResult to use the synchronous broadcast method and return the value of a boolean extra property HANDLED from the Intent:

```
private boolean broadcastResult(String result) {
    Intent intent = new Intent(PRIMES_BROADCAST);
    intent.putExtra(RESULT, result);
    LocalBroadcastManager.getInstance(this).
      sendBroadcastSync(intent);
    return intent.getBooleanExtra(HANDLED, false);
}
```

Because sendBroadcastSync is synchronous, all registered BroadcastReceivers will have handled the broadcast by the time sendBroadcastSync returns. This means that if any receiver sets the Boolean "extra" property HANDLED to true, broadcastResult will return true.

In our BroadcastReceiver, we'll update the Intent object by adding a boolean property to indicate that we've handled it:

```
@Override
public void onReceive(Context context, Intent intent) {
    if (view != null) {
        String result = intent.getStringExtra(
          PrimesServiceWithBroadcast.RESULT);
        intent.putExtra(LocalPrimesService.HANDLED, true);
        view.setText(result);
```

```
  } else {
    Log.i(TAG, " ignoring - we're detached");
  }
}
```

Now if `PrimesActivity` is still running, its `BroadcastReceiver` is registered and receives the `Intent` object and will put the extra `boolean` property `HANDLED` with the value `true`.

However, if `PrimesActivity` has finished, the `BroadcastReceiver` will no longer be registered and `LocalPrimesService` will return `false` from its `broadcastResult` method. We can use this to decide whether we should post a notification:

```
if (!broadcastResult(prime.toString()))
  notifyUser(n, prime.toString());
```

There's one final complication: unlike `sendBroadcast`, which always invokes BroadcastReceivers on the main thread, `sendBroadcastSync` uses the thread that it is called with. Our `BroadcastReceiver` interacts directly with the user interface, so we must call it on the main thread. This is simple enough to achieve from our `AsyncTask` — we'll move the broadcast from `doInBackground` to `onPostExecute`:

```
public void calculateNthPrime(final int n) {
  new AsyncTask<Void,Void,BigInteger>(){
    @Override
    protected BigInteger doInBackground(Void... params) {
      BigInteger prime = new BigInteger("2");
      for (int i=0; i<n; i++)
        prime = prime.nextProbablePrime();
      return prime;
    }
    @Override
    protected void onPostExecute(BigInteger result) {
      if (!broadcastResult(result.toString()))
        notifyUser(n, result.toString());
    }
  }.execute();
}
```

This does just what we want — if our `BroadcastReceiver` handles the message, we don't post a notification; otherwise, we will do so to make sure the user gets their result.

Having developed a good understanding of how to use Services to conduct long-running background work, let's consider some real-world applications and use cases.

Applications of Services

With a little bit of work, Services give us the means to perform long-running background tasks, and free us from the tyranny of the `Activity` lifecycle. Unlike `IntentService`, directly subclassing `Service` also gives us the ability control the level of concurrency.

With the ability to run as many tasks as we need and to take as long as is necessary to complete those tasks, a world of new possibilities opens up.

The only real constraint on how and when we use Services comes from the need to communicate results to a user-interface component, such as a `Fragment` or `Activity`, and the complexity this entails.

Ideal use cases for Services tend to have the following characteristics:

- Long-running (a few hundred milliseconds and upward)
- Not specific to a single `Activity` or `Fragment` class
- Must complete, even if the user leaves the application
- Requires more concurrency than `IntentService` provides, or needs control over the level of concurrency

There are many applications that exhibit these characteristics, but the stand-out example is, of course, handling concurrent downloads from a web service.

To make good use of the available download bandwidth and to limit the impact of network latency, we want to be able to run more than one download at a time (but not too many). We also don't want to use more bandwidth than necessary by failing to completely download a file and having to restart the download later. So ideally, once a download starts, it should run to completion even if the user leaves the application.

In the accompanying downloads from the Packt Publishing website, you can find a complete example of using a `Service` to download NASA's "Image of the Day" RSS, parse the XML in the background to extract the titles and URLs of the images, and download and display those images.

Summary

In this chapter, we explored the very powerful Service component, putting it to use to execute long-running background tasks with a configurable level of concurrency.

We learned that running tasks in a Service gives them the best possible chance of successful completion if the user exits the application, because the system avoids killing processes with active Services unless absolutely necessary.

We discovered various ways to initiate background work in a Service, from starting the Service with an Intent object through to direct method invocation using local Services.

We also saw the wide range of communication mechanisms available for delivering results back to the user: direct invocation of local Service methods; sending Messages with Messenger; broadcasting results to registered parties with BroadcastReceiver; and, if the user has already left the application, raising system notifications.

Of course, we can also use PendingIntent to send data back to the originating Activity, just as we did with IntentService in *Chapter 5, Queuing Work with IntentService.*

In the next chapter, we'll add one last weapon to our arsenal: the ability to run background tasks at specific times—even when the device is asleep—by scheduling alarms with AlarmManager.

7
Scheduling Work with AlarmManager

Maintaining the responsiveness of foreground apps has been our primary focus throughout this book, and we've explored numerous ways to shift work away from the main thread to run in the background.

In all of our discussions so far, we wanted to get the work done as soon as possible, so although we moved it off to a background thread, we still performed the work concurrently with ongoing main thread operations, such as updating the user interface and responding to user interaction.

In this chapter we'll learn to schedule work to run at some distant time in the future, launching our application without user intervention if it isn't already running, and even waking the device from sleep if necessary.

In this chapter we will cover:

- Scheduling alarms with AlarmManager
- Canceling alarms
- Scheduling repeating alarms
- Handling alarms
- Staying awake with WakeLocks
- Applications of AlarmManager

Scheduling alarms with AlarmManager

In *Chapter 3, Distributing Work with Handler and HandlerThread*, we learned to schedule work on a `HandlerThread` using `postDelayed`, `postAtTime`, `sendMessageDelayed`, and `sendMessageAtTime`. These mechanisms are fine for short-term scheduling of work that should happen soon—while our application is running in the foreground.

However, if we want to schedule an operation to run at some point in the distant future, we'll run into problems. First, our application may be terminated before that time arrives, removing any chance of the `Handler` running those scheduled operations. Second, the device may be asleep, and with its CPU powered down, it cannot run our scheduled tasks.

The solution to this is to use an alternative scheduling approach—one that is designed to overcome these problems: `AlarmManager`.

`AlarmManager` is a system service that provides scheduling capabilities far beyond those of `Handler`. Being a system service, `AlarmManager` cannot be terminated and has the capacity to wake the device from sleep to deliver scheduled alarms.

We can access `AlarmManager` via a `Context` instance, so from any lifecycle callback in an `Activity`, we can get the `AlarmManager` by using the following code:

```
AlarmManager am = (AlarmManager)getSystemService(ALARM_SERVICE);
```

Once we have a reference to the `AlarmManager`, we can schedule an alarm to deliver a `PendingIntent` object at a time of our choosing. The simplest way to do that is using the `set` method.

```
void set(int type, long triggerAtMillis, PendingIntent operation)
```

When we set an alarm we must also specify a `type` flag—the first parameter to the `set` method. The `type` flag sets the conditions under which the alarm should fire and which `clock` to use for our schedule.

There are two conditions and two clocks, resulting in four possible `type` settings.

The conditions specify whether or not the device will be woken up if it is sleeping at the time of the scheduled alarm—whether the alarm is a "wakeup" alarm or not.

The clocks provide a reference time against which we set our schedules, defining exactly what we mean when we set a value to `triggerAtMillis`.

The elapsed-time system clock — `android.os.SystemClock` — measures time as the number of milliseconds that have passed since the device booted, including any time spent in deep sleep. The current time according to the system clock can be found using:

```
SystemClock.elapsedRealtime()
```

The real-time clock measures time in milliseconds since the Unix epoch. The current time according to the real-time clock can be found with:

```
System.currentTimeMillis()
```

> In Java, `System.currentTimeMillis()` returns the number of milliseconds since midnight on January 1, 1970, **Coordinated Universal Time (UTC)** — a point in time known as the **Unix epoch**.
>
> UTC is the internationally recognized successor to **Greenwich Mean Time (GMT)** and forms the basis for expressing international time zones, which are typically defined as positive or negative offsets from UTC.

Given these two conditions and two clocks, the four possible `type` values we can use when setting alarms are:

- `AlarmManager.ELAPSED_REALTIME`: This schedules the alarm relative to the system clock. If the device is asleep at the scheduled time it will not be delivered immediately, instead the alarm will be delivered the next time the device wakes.

- `AlarmManager.ELAPSED_REALTIME_WAKEUP`: This schedules the alarm relative to the system clock. If the device is asleep, it will be woken to deliver the alarm at the scheduled time.

- `AlarmManager.RTC`: This schedules the alarm in UTC relative to the Unix epoch. If the device is asleep at the scheduled time, the alarm will be delivered when the device is next woken.

- `AlarmManager.RTC_WAKEUP`: This schedules the alarm relative to the Unix epoch. If the device is asleep it will be awoken, and the alarm is delivered at the scheduled time.

Let's consider a few examples. We'll use the `TimeUnit` class from the `java.lang.`
`concurrent` package to calculate times in milliseconds. To set an alarm to go off
48 hours after the initial boot, we need to work with the system clock, as shown
in the following code:

```
long delay = TimeUnit.HOURS.toMillis(48L);
long time = System.currentTimeMillis() + delay;
am.set(AlarmManager.ELAPSED_REALTIME, time, pending);
```

We can set an alarm to go off in 2 hours from now using a clock, by adding two hours
to the current time. Using the system clock it looks like this:

```
long delay = TimeUnit.HOURS.toMillis(2L);
long time = SystemClock.elapsedRealtime() + delay;
am.set(AlarmManager.ELAPSED_REALTIME, time, pending);
```

To set an alarm for 2 hours from now using the real-time clock is similar to the
previous code:

```
long delay = TimeUnit.HOURS.toMillis(2L);
long time = System.currentTimeMillis() + delay;
am.set(AlarmManager.RTC, time, pending);
```

To set an alarm for 9 p.m. today (or tomorrow, if it's already past 9 p.m. today):

```
Calendar calendar = Calendar.getInstance();
if (calendar.get(Calendar.HOUR_OF_DAY) >= 21) {
  calendar.add(Calendar.DATE, 1);
}
calendar.set(Calendar.HOUR_OF_DAY, 21);
calendar.set(Calendar.MINUTE, 0);
calendar.set(Calendar.SECOND, 0);
am.set(AlarmManager.RTC, calendar.getTimeInMillis(), pending);
```

None of the examples so far will wake the device if it is sleeping at the time of the
alarm. To do that we need to use one of the `WAKEUP` alarm conditions, for example:

```
am.set(AlarmManager.ELAPSED_REALTIME_WAKEUP, time, pending);
am.set(AlarmManager.RTC_WAKEUP, time, pending);
```

If our application targets an API level below 19 (KitKat), alarms scheduled with
`set` will run at exactly the alarm time. For applications targeting KitKat or greater,
the schedule is considered inexact and the system may re-order or group alarms
to minimize wake-ups and save battery.

If we need precision scheduling and are targeting KitKat or greater, we can use the new `setExact` method introduced at API level 19, but we'll need to check that the method exists before we try to call it; otherwise, our app will crash when run under earlier API levels:

```
if (Build.VERSION.SDK_INT >= 19) {
    am.setExact(AlarmManager.RTC_WAKEUP, time pending);
} else {
    am.set(AlarmManager.RTC_WAKEUP, time, pending);
}
```

This will deliver our alarm at exactly the specified time on all platforms. We should only use exact scheduling when we really need it—for example, to deliver alerts to the user at a specific time. For most other cases, allowing the system to adjust our schedule a little to preserve battery life is usually acceptable.

One more addition in KitKat is `setWindow`, which introduces a compromise between exact and inexact alarms by allowing us to specify the time window within which the alarm must be delivered. This still allows the system some freedom to play with the schedules for efficiency, but lets us choose just how much freedom to allow.

Here's how we would use `setWindow` to schedule an alarm to be delivered within a 3 minute window—at the earliest 10 minutes from now and at the latest 13 minutes from now—using the real-time clock:

```
long delay = TimeUnit.MINUTES.toMillis(10);
long window = TimeUnit.MINUTES.toMillis(3);
long earliest = System.currentTimeMillis() + delay;
am.setWindow(
    AlarmManager.RTC_WAKEUP, earliest, window, pending);
```

Canceling alarms

Once set, an alarm can be canceled very easily—we just need to invoke the AlarmManager's `cancel` method with an `Intent` matching that of the alarm we want to cancel.

The process of matching uses the `filterEquals` method of `Intent`, which compares the action, data, type, class, and categories of both Intents to test for equivalence. Any extras we may have set in the `Intent` are not taken into account. We can set and cancel an alarm using different `Intent` instances like this:

```
public void setThenCancel() {
    AlarmManager am = (AlarmManager)
```

```
                    getSystemService(ALARM_SERVICE);
        long at =
            System.currentTimeMillis() +
            TimeUnit.SECONDS.toMillis(5L);
        am.set(AlarmManager.RTC, at, createPendingIntent());
        am.cancel(createPendingIntent());
    }

    private PendingIntent createPendingIntent() {
        Intent intent = new Intent("my_action");
        // extras don't affect matching
        intent.putExtra("random", Math.random());
        return PendingIntent.getBroadcast(
            this, 0, intent, PendingIntent.FLAG_UPDATE_CURRENT);
    }
```

It is important to realize that whenever we set an alarm, we implicitly cancel any existing alarm with a matching Intent, replacing it with the new schedule.

Scheduling repeating alarms

As well as setting a one-off alarm, we have the option to schedule repeating alarms using setRepeating and setInexactRepeating. Both methods take an additional parameter that defines the interval in milliseconds at which to repeat the alarm. Generally it is advisable to avoid setRepeating and always use setInexactRepeating, allowing the system to optimize wake-ups and giving more consistent behavior on devices running different Android versions:

```
void setRepeating(
    int type, long triggerAtMillis,
    long intervalMillis, PendingIntent operation);
void setInexactRepeating(
    int type, long triggerAtMillis,
    long intervalMillis, PendingIntent operation)
```

AlarmManager provides some handy constants for typical repeat intervals:

```
AlarmManager.INTERVAL_FIFTEEN_MINUTES
AlarmManager.INTERVAL_HALF_HOUR
AlarmManager.INTERVAL_HOUR
AlarmManager.INTERVAL_HALF_DAY
AlarmManager.INTERVAL_DAY
```

We can schedule a repeating alarm to be delivered approximately 2 hours from now, then repeating every 15 minutes or so thereafter like this:

```
Intent intent = new Intent("my_action");
PendingIntent broadcast = PendingIntent.getBroadcast(
    this, 0, intent, PendingIntent.FLAG_UPDATE_CURRENT);
long start =
    System.currentTimeMillis() +
    TimeUnit.HOURS.toMillis(2L);
AlarmManager am = (AlarmManager)
    getSystemService(ALARM_SERVICE);
am.setRepeating(
    AlarmManager.RTC_WAKEUP, start,
    AlarmManager.INTERVAL_FIFTEEN_MINUTES, broadcast);
```

From API level 19, all repeating alarms are inexact—that is, if our application targets KitKat or above, our repeat alarms will be inexact even if we use `setRepeating`.

If we really need exact repeat alarms, we can use `setExact` instead, and schedule the next alarm while handling the current one.

Handling alarms

Now that we know how to schedule alarms, let's take a look at what we can schedule with them.

Essentially, we can schedule anything that can be started with a `PendingIntent`, which means we can use alarms to start Activities, Services, and BroadcastReceivers. To specify the target of our alarm, we need to use the static factory methods of `PendingIntent`:

```
PendingIntent.getActivities(...)
PendingIntent.getActivity(...)
PendingIntent.getService(...)
PendingIntent.getBroadcast(...)
```

In the following sections, we'll see how each type of `PendingIntent` can be used with `AlarmManager`.

Handling alarms with Activities

Starting an `Activity` from an alarm is as simple as registering the alarm with a `PendingIntent` created by invoking the static `getActivity` method of `PendingIntent`.

When the alarm is delivered, the `Activity` will be started and brought to the foreground, displacing any app that was currently in use. Keep in mind that this is likely to surprise and perhaps annoy users!

When starting Activities with alarms, we will probably want to set `Intent.FLAG_ACTIVITY_CLEAR_TOP`; so that if the application is already running, our target `Activity` assumes its normal place in the navigation order within the app.

```
Intent intent = new Intent(context, HomeActivity.class);
intent.setFlags(Intent.FLAG_ACTIVITY_CLEAR_TOP);
PendingIntent pending = PendingIntent.getActivity(
    Context, 0, intent, PendingIntent.FLAG_UPDATE_CURRENT);
```

Not all Activities are suited to being started with `getActivity`. We might need to start an `Activity` that normally appears deep within the app, where pressing "back" does not exit to the home screen, but returns to the next `Activity` on the back-stack.

This is where `getActivities` comes in. With `getActivities`, we can push more than one `Activity` onto the back-stack of the application, allowing us to populate the back-stack to create the desired navigation flow when the user presses "back". To do this, we create our `PendingIntent` by sending an array of Intents to `getActivities`:

```
Intent home = new Intent(context, HomeActivity.class);
Intent target = new Intent(context, TargetActivity.class);
home.setFlags(Intent.FLAG_ACTIVITY_CLEAR_TOP);
PendingIntent pending = PendingIntent.getActivities(
    context, 0, new Intent[]{ home, target },
    PendingIntent.FLAG_UPDATE_CURRENT);
```

The array of Intents specifies the Activities to launch, in order. The logical sequence of events when this alarm is delivered is:

1. If the application is already running, any Activities on the back-stack above `HomeActivity` are finished and removed, because we set the `Intent.FLAG_ACTIVITY_CLEAR_TOP` flag.

2. `HomeActivity` is (re)started.

3. `TargetActivity` is started and placed on the back-stack above `HomeActivity`. The `TargetActivity` becomes the foreground `Activity`

Handling alarms with Activities is good to know about, but is not a technique we will use often, since it is so intrusive. We are much more likely to want to handle alarms in the background, which we'll look at next.

Handling alarms with BroadcastReceiver

We met BroadcastReceiver already in *Chapter 6, Long-running Tasks with Service*, where we used it in an Activity to receive broadcasts from an IntentService. In this section, we'll use BroadcastReceiver to handle alarms.

BroadcastReceivers can be registered and unregistered dynamically at runtime — like we did in *Chapter 6, Long-running Tasks with Service*, or statically in the Android manifest file with a <receiver> element, and can receive alarms regardless of how they are registered.

It is more common to use a statically registered receiver for alarms, because these are known to the system and can be invoked by alarms to start an application if it is not currently running. A receiver registration in the manifest looks like this:

```
<receiver android:name=".AlarmReceiver">
    <intent-filter>
        <action android:name="reminder"/>
    </intent-filter>
</receiver>
```

The <intent-filter> element gives us the opportunity to say which Intents we want to receive, by specifying the action, data, and categories that should match. Here we're just matching Intents with the action reminder, so we can set an alarm for this receiver with:

```
Intent intent = new Intent("reminder");
intent.putExtra(AlarmReceiver.MSG, "try the alarm examples!");
PendingIntent broadcast = PendingIntent.getBroadcast(
    context, 0, intent, PendingIntent.FLAG_UPDATE_CURRENT);
AlarmManager am = (AlarmManager) getSystemService(ALARM_SERVICE);
long at = calculateNextMidnightRTC();
am.set(AlarmManager.RTC_WAKEUP, at, broadcast);
```

When this alarm comes due, AlarmManager will wake the device — if it isn't already awake — and deliver the Intent to the BroadcastReceiver's onReceive method.

```
public class AlarmReceiver extends BroadcastReceiver {
    public static final String MSG = "message";
    @Override
```

```
        public void onReceive(Context ctx, Intent intent) {
            // do some work while the device is awake
        }
    }
```

The `AlarmManager` guarantees that the device will remain awake at least until `onReceive` completes, which means we can be sure of getting some work done before the device will be allowed to return to sleep.

Doing work with BroadcastReceiver

When the system delivers an alarm to our `BroadcastReceiver` it does so on the main thread, so the usual main thread limitations apply: we cannot perform networking and we should not perform heavy processing or use blocking operations.

In addition, a statically registered `BroadcastReceiver` has a very limited lifecycle. It cannot create user-interface elements other than toasts or notifications posted via `NotificationManager`; the `onReceive` method must complete within 10 seconds or its process may be killed; and once `onReceive` completes, the receiver's life is over.

If the work that we need to do in response to the alarm is not intensive, we can simply complete it during `onReceive`. A good use for BroadcastReceivers that receive alarms is posting notifications to the notification drawer as follows:

```
public class AlarmReceiver extends BroadcastReceiver {
    public static final String MSG = "message";
    @Override
    public void onReceive(Context ctx, Intent intent) {
        NotificationCompat.Builder builder =
            new NotificationCompat.Builder(context)
                .setSmallIcon(android.R.drawable.stat_notify_chat)
                .setContentTitle("Reminder!")
                .setContentText(intent.getStringExtra(MSG));
        NotificationManager nm = (NotificationManager)
            context.getSystemService(
                Context.NOTIFICATION_SERVICE);
        nm.notify(intent.hashCode(), builder.build());
    }
}
```

We can make this more useful by including a `PendingIntent` object in the notification, so that the user can tap the notification to open an `Activity` in our app:

```
@Override
    public void onReceive(Context ctx, Intent intent) {
        Intent activity = new Intent(
```

```
        ctx, HomeActivity.class);
    intent.setFlags(Intent.FLAG_ACTIVITY_NEW_TASK);
    PendingIntent pending = PendingIntent.getActivity(
        ctx, 0, activity,
        PendingIntent.FLAG_UPDATE_CURRENT);
    NotificationCompat.Builder builder =
        new NotificationCompat.Builder(context)
            .setSmallIcon(android.R.drawable.stat_notify_chat)
            .setContentTitle("Reminder!")
            .setContentText(intent.getStringExtra(MSG))
            .setContentIntent(pending)
            .setAutoCancel(true);
    NotificationManager nm = (NotificationManager)
        context.getSystemService(
            Context.NOTIFICATION_SERVICE);
    nm.notify(intent.hashCode(), builder.build());
}
```

Although we can spend up to 10 seconds doing work in our BroadcastReceiver, we really shouldn't—if the app is in use when the alarm is triggered the user will suffer noticeable lag if onReceive takes more than a hundred milliseconds to complete. Exceeding the 10 second budget will cause the system to kill the application and report a background ANR.

A second option is available to us only if we're targeting API level 11 and above, and allows onReceive to delegate work to a background thread for up to 10 seconds—we'll discuss this in the next section.

Doing background work with goAsync

If our application targets a minimum API level of 11, we can use a feature of BroadcastReceiver that it introduced: goAsync.

```
    public final PendingResult goAsync()
```

With goAsync we can extend the lifetime of a BroadcastReceiver instance beyond the completion of its onReceive method, provided the whole operation still completes within the 10 second budget.

If we invoke goAsync, the system will not consider the BroadcastReceiver to have finished when onReceive completes. Instead, the BroadcastReceiver lives on until we call finish on the PendingResult returned to us by goAsync. We must ensure that finish is called within the 10 second budget, otherwise the system will kill the process with a background ANR.

Using `goAsync` we can offload work to background threads using any appropriate concurrency construct—for example, an `AsyncTask`—and the device is guaranteed to remain awake until we call `finish` on the `PendingResult`.

```
public void onReceive(final Context context, final Intent intent) {
    final PendingResult result = goAsync();
    new AsyncTask<Void,Void,Void>(){
        @Override
        protected Void doInBackground(Void... params) {
            try {
                // … do some work here, for up to 10 seconds
            } finally {
                result.finish();
            }
        }
    }.execute();
}
```

This is nice; though its utility is limited by the 10 second budget and the effects of fragmentation (it is only available to API level 11 or above). In the next section, we'll look at scheduling long-running operations with services.

Handling alarms with Services

Just like starting Activities, starting a `Service` from an alarm involves scheduling an appropriate `PendingIntent` instance, this time using the static `getService` method.

We almost certainly want our `Service` to do its work off the main thread, so sending work to an `IntentService` this way seems ideal, and an `IntentService` will also stop itself when the work is finished.

This works beautifully if the device is awake.

However, if the device is asleep we have a potential problem. AlarmManager's documentation tells us that the only guarantee we have about the wakefulness of the device is that it will remain awake until a BroadcastReceiver's `onReceive` method completes.

Since directly starting a `Service` does not involve a `BroadcastReceiver`, and in any case is an asynchronous operation, there is no guarantee that the `Service` will have started up before the device returns to sleep, so the work may not get done until the device is next awakened.

This is almost certainly not the behavior we want. We want to ensure that the Service starts up and completes its work, regardless of whether the device was awake when the alarm was delivered. To do that, we'll need a BroadcastReceiver and a little explicit power management, as we'll see next.

Staying awake with WakeLocks

Earlier in this chapter we learned that we can use a BroadcastReceiver to handle alarms, and even do work in the background for up to 10 seconds, though only on devices running API level 11 or greater.

In the previous section, we saw that handling alarms directly with services is not a reliable solution for scheduling long-running work, since there is no guarantee that our Service will start up before the device returns to sleep.

We have a problem! If we want to perform long-running work in response to alarms, we need a solution that overcomes these limitations.

What we really want is to start a Service to handle the work in the background, and to keep the device awake until the Service has finished its work. Fortunately, we can do that by combining the wakefulness guarantees of BroadcastReceiver to get the Service started, then keep the device awake with explicit power management using PowerManager and WakeLock.

As you might guess, WakeLock is a way to force the device to stay awake. WakeLocks come in various flavors, allowing apps to keep the screen on at varying brightness levels or just to keep the CPU powered up in order to do background work. To use WakeLocks, our application must request an additional permission in the manifest:

```
<uses-permission android:name="android.permission.WAKE_LOCK" />
```

To keep the CPU powered up while we do background work in a Service, we only need a partial WakeLock, which won't keep the screen on, and which we can request from the PowerManager like this:

```
PowerManager pm = (PowerManager)
    ctx.getSystemService(Context.POWER_SERVICE);
WakeLock lock = pm.newWakeLock(
    PowerManager.PARTIAL_WAKE_LOCK, "my_app");
lock.acquire();
```

We'll need to acquire a WakeLock during our BroadcastReceiver's onReceive method, and find some way to hand it to our Service so that the Service can release the lock once its work is done.

Unfortunately, WakeLocks are not parcelable, so we can't just send them to the Service in an Intent. The simplest solution is to manage the WakeLock instance as a static property that both the BroadcastReceiver and the target Service can reach.

This is not difficult to implement, but we don't actually need to implement it ourselves — we can use the handy v4 support library class, WakefulBroadcastReceiver.

WakefulBroadcastReceiver exposes two static methods that take care of acquiring and releasing a partial WakeLock. We can acquire the WakeLock, and start the Service with a single call to startWakefulService:

```
ComponentName startWakefulService(Context context, Intent intent);
```

And when our Service has finished its work, it can release the WakeLock with the corresponding call to completeWakefulIntent:

```
boolean completeWakefulIntent(Intent intent);
```

Since these methods are static, we don't need to extend WakefulBroadcastReceiver to use them. Our BroadcastReceiver just needs to start the Service using startWakefulService:

```
class AlarmReceiver extends BroadcastReceiver {
    @Override
    public void onReceive(Context context, Intent intent) {
        Intent serviceIntent = new Intent(
            context, MyIntentService.class);
        WakefulBroadcastReceiver.startWakefulService(
            context, serviceIntent);
    }
}
```

We must make sure to release the WakeLock once the Service has finished its work, otherwise we'll drain the battery by keeping the CPU powered up unnecessarily:

```
class MyIntentService extends IntentService {
    @Override
    protected final void onHandleIntent(Intent intent) {
        try {
            // do background work while the CPU is kept awake
        } finally {
            WakefulBroadcastReceiver.
                completeWakefulIntent(intent);
        }
    }
}
```

This is great—by using a statically registered BroadcastReceiver we've ensured that we receive the alarm, even if our application is not running when the alarm comes due. When we receive the alarm we acquire a WakeLock, keeping the device awake while our Service starts up and does its potentially long-running work. Once our work is done, we release the WakeLock to allow the device to sleep again and conserve power.

Applications of AlarmManager

AlarmManager allows us to schedule work to run without user intervention. This means that we can arrange to do work pre-emptively, for example, to prepare data that our application will need to present to the user when they next open the application, or to alert the user to new or updated information with notifications.

Ideal use cases include things like periodically checking for new e-mails, downloading new editions of periodical publications (for example, daily newspapers), or uploading data from the device to a cloud backup service.

Summary

In this chapter we learned to schedule work for our applications to perform at some time in the future, either as a one-off operation or at regular intervals.

We learned to set alarms relative to the system clock or real time, how to wake the device up from a deep sleep, and how to cancel alarms when we no longer need them.

Our exploration covered various options for responding to alarms, including bringing an Activity to the foreground or doing work directly in a BroadcastReceiver, synchronously or asynchronously.

Finally, we arranged for a Service to be launched with a WakeLock to keep the CPU from powering down while our long-running background work is completed.

Over the course of this book, we've armed ourselves with a powerful array of tools for building responsive Android applications. We discovered that it is incredibly important to shift as much work as possible off the main thread, and explored a number of constructs and techniques for doing that.

We now know how to move short-term operations to background threads using `AsyncTask` and `HandlerThread`. We learned how to perform asynchronous I/O to load data from local files and databases with `Loader`, and then discovered how, when, and why to use `Service` and its more specialized subclass `IntentService` to give our long-running tasks the best possible chance of being completed successfully. Finally, we learned how to schedule tasks to run at some future time using `AlarmManager`.

By keeping in mind that the main thread is sacred, and using the available constructs to perform work on background threads whenever possible, our applications will be smooth and responsive, and have the best possible chance of delighting our users.

Index

M

MediaCursorAdapter 68
mutual exclusion 12

O

onCanceled method 62
onCancelled callback method 17
onContentChanged method 70
onCreateLoader method 58
onHandleIntent method 93
onLoaderFinished method 58
onLoaderReset method 58
onLoadFinished method 70
onPreExecute method 16
onProgressUpdate callback method 22
onStartCommand method 93
onStartLoading method 70
onStopLoading method 62

P

PendingIntent
  about 117
  static factory methods 117
PrimesIntentService 98

R

race condition 12
readLine method 48
responsive apps
  building, AsyncTaskLoader used 59-66
  building, CursorLoader used 66-69
  building, with Handler 38
  building, with Service 92
  builsing, with HandlerThread 50, 51
run() method 12

S

Service
  about 76
  alarms, handling 122
  applications 108
  communicating with 98

direct communication, with
    local Services 99-102
  Executor interface, controlling 92-96
  results, broadcasting with Intents 103-105
  results, returning with Messenger 96-98
  unhandled broadcasts, detecting 106, 107
  used, for building responsive apps 92
SimpleCursorAdapter class 68
SimpleLooper class 36
start() method 12

T

ThumbnailLoader
  enabling 69

W

WakeLocks
  about 123
  used, for staying awake 123-125
work, posting to Handler
  pending Runnable, cancelling 42
  performing 40, 41
  Runnable, executing 39
work, posting with send
  composition, versus inheritance 45, 46
  multithreaded example 47-49
  pending Messages, cancelling 45
  performing 42-44

Z

Zygote
  about 9
  benefits 9

About Packt Publishing

Packt, pronounced 'packed', published its first book "*Mastering phpMyAdmin for Effective MySQL Management*" in April 2004 and subsequently continued to specialize in publishing highly focused books on specific technologies and solutions.

Our books and publications share the experiences of your fellow IT professionals in adapting and customizing today's systems, applications, and frameworks. Our solution based books give you the knowledge and power to customize the software and technologies you're using to get the job done. Packt books are more specific and less general than the IT books you have seen in the past. Our unique business model allows us to bring you more focused information, giving you more of what you need to know, and less of what you don't.

Packt is a modern, yet unique publishing company, which focuses on producing quality, cutting-edge books for communities of developers, administrators, and newbies alike. For more information, please visit our website: www.packtpub.com.

About Packt Open Source

In 2010, Packt launched two new brands, Packt Open Source and Packt Enterprise, in order to continue its focus on specialization. This book is part of the Packt Open Source brand, home to books published on software built around Open Source licences, and offering information to anybody from advanced developers to budding web designers. The Open Source brand also runs Packt's Open Source Royalty Scheme, by which Packt gives a royalty to each Open Source project about whose software a book is sold.

Writing for Packt

We welcome all inquiries from people who are interested in authoring. Book proposals should be sent to author@packtpub.com. If your book idea is still at an early stage and you would like to discuss it first before writing a formal book proposal, contact us; one of our commissioning editors will get in touch with you.

We're not just looking for published authors; if you have strong technical skills but no writing experience, our experienced editors can help you develop a writing career, or simply get some additional reward for your expertise.

Augmented Reality for Android Application Development

ISBN: 978-1-78216-855-3   Paperback: 130 pages

Learn how to develop advanced Augmented Reality applications for Android

1. Understand the main concepts and architectural components of an AR application

2. Step-by-step learning through hands-on programming combined with a background of important mathematical concepts

3. Efficiently and robustly implement some of the main functional AR aspects

Android Studio Application Development

ISBN: 978-1-78328-527-3   Paperback: 110 pages

Create visually appealing applications using the new IntelliJ IDE Android Studio

1. Familiarize yourself with Android Studio IDE

2. Enhance the user interface for your app using the graphical editor feature

3. Explore the various features involved in developing an android app and implement them

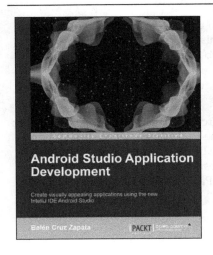

Please check **www.PacktPub.com** for information on our titles

Android Native Development Kit Cookbook

ISBN: 978-1-84969-150-5   Paperback: 346 pages

A step-by-step tutorial with more than 60 concise recipes on Android NDK development skills

1. Build, debug, and profile Android NDK apps

2. Implement part of Android apps in native C/C++ code

3. Optimize code performance in assembly with Android NDK

Android Development Tools for Eclipse

ISBN: 978-1-78216-110-3   Paperback: 144 pages

Set up, build, and publish Android projects quickly using Android Development Tools for Eclipse

1. Build Android applications using ADT for Eclipse

2. Generate Android application skeleton code using wizards

3. Advertise and monetize your applications

Please check **www.PacktPub.com** for information on our titles